79 Springer Series in Solid-State Sciences

Edited by Peter Fulde

Springer Series in Solid-State Sciences

Editors: M. Cardona   P. Fulde   K. von Klitzing   H.-J. Queisser

Managing Editor: H. K. V. Lotsch

K. Ohbayashi   M. Watabe (Eds.)

Elementary Excitations in Quantum Fluids

Proceedings of the Hiroshima Symposium,
Hiroshima, Japan, August, 17–18, 1987

With 44 Figures

Springer-Verlag Berlin Heidelberg New York
London Paris Tokyo

Professor Kohji Ohbayashi
Professor Mitsuo Watabe

Faculty of Integrated Arts and Sciences,
Hiroshima University, Hiroshima 730, Japan

Series Editors:

Professor Dr., Dres. h. c. Manuel Cardona
Professor Dr., Dr. h. c. Peter Fulde
Professor Dr. Klaus von Klitzing
Professor Dr. Hans-Joachim Queisser

Max-Planck-Institut für Festkörperforschung, Heisenbergstrasse 1
D-7000 Stuttgart 80, Fed. Rep. of Germany

Managing Editor: Dr. Helmut K. V. Lotsch

Springer-Verlag, Tiergartenstrasse 17, D-6900 Heidelberg, Fed. Rep. of Germany

ISBN 3-540-19106-2 Springer-Verlag Berlin Heidelberg New York
ISBN 0-387-19106-2 Springer-Verlag New York Berlin Heidelberg

The use of registered names, trademarks, etc. in this publication does not imply, even in the absence of a specific statement, that such names are exempt from the relevant protective laws and regulations and therefore free for general use.

Printing: Druckhaus Beltz, 6944 Hemsbach/Bergstr.; Binding: J. Schäffer GmbH & Co. KG., 6718 Grünstadt
2154/3150-543210 – Printed on acid-free paper

Preface

This volume is the proceedings of the Hiroshima Symposium on Elementary Excitations in Quantum Fluids, which was held on August 17 and 18, 1987, in Hiroshima, Japan, and was attended by thirty-two scientists from seven countries.

Quantum fluids have been the subject of intense study as a consequence of their superfluid properties at very low temperatures. Elementary excitations in them are an important concept about which many important discoveries have been made in recent years. This symposium was arranged by a group of physicists from Hiroshima University to provide an opportunity to discuss these recent developments. It was conceived as a satellite conference of the 18th International Conference on Low Temperature Physics (LT 18), which was held in Kyoto, August 20–26, 1987.

Emphasis was placed on the dynamic structures and correlations of elementary excitations, which resulted in invited speakers being selected from this field. However, enthusiastic contributors reported notable new results on various other aspects of the elementary excitations, which made the symposium lively and successful. It is our great satisfaction to present this volume, which includes papers of good quality and originality. We thank all the participants for their cooperation throughout this symposium, and for preparing their manuscripts within a reasonable time.

Obviously, without the sponsors there would have been no Hiroshima Symposium. Therefore, it is our pleasure to acknowledge in the name of the participants those organizations who contributed to our symposium: The Ministry of Education, Science, and Culture of Japan; Hitachi, Ltd.; Kobe Steel, Ltd.; Japan Spectroscopic Co., Ltd.; Kowa Company, Ltd.; Columbia Import & Export Co., Ltd.; NEC Corporation; and the Rare Metallic Co., Ltd.

A word of thanks is due to the members of the Faculty of Integrated Arts and Sciences, Hiroshima University, especially Dr. M. Udagawa, who collaborated in arranging the symposium as well as guiding the tour to the beautiful Miyajima Island.

We would like to take this opportunity to extend our thanks to Professor R. Kubo and Professor S. Nakajima who have continuously supported our group. In particular, as the chairman of LT 18, Professor Nakajima encouraged us to hold this symposium.

Hiroshima, May 1988

Kohji Ohbayashi
Mitsuo Watabe

The arch-shaped monument (32) is the Peace Memorial Cenotaph in the Peace Memorial Park, on which is inscribed a sentence:

Let All the Souls Here Rest in Peace; For We Shall Not Repeat the Evil.

(Photo by Y. Utunomiya)

Hiroshima Symposium on **Elementary Excitations in Quantum Fluids**
August 17–18, 1987

1. J. W. Halley	12. V. L. P. Frank	23. K. Fukushima
2. S.Nakajima	13. K. Ishikawa	24. C. D. H. Williams
3. C.E. Campbell	14. E. C. Svensson	25. S. Komura
4. K. Ohbayashi	15. H. J. Lauter	26. K. F. Coates
5. A. Thellung	16. N. Ogita	27. S. Sasaki
6. J. Ruvalds	17. A. F. G. Wyatt	28. H. Yoshiki
7. K. Nagai	18. M. Udagawa	29. M. Hasegawa
8. A. Fukumoto	19. R. Golub	30. W. F. Vinen
9. A. Griffin	20. G. Wyborn	31. P. V. M. McClintock
10. M. Watabe	21. G. A. Williams	
11. F. Iwamoto	22. D. Hirashima	

Contents

Part IV   **Quantum Evaporation**

Part V   **Vortices, Ripplons and Ions**

Introduction

K. Ohbayashi and M. Watabe

Faculty of Integrated Arts and Sciences, Hiroshima University,
Hiroshima 730, Japan

This symposium was designed to report and to discuss recent research into excitations in quantum fluids such as phonons, rotons, maxons, plateaus, vortices, ripplons, and local modes in films, mainly in superfluid ^{4}He (He II). Many new interesting experimental results were reported, which were obtained by various methods such as neutron scattering, Raman scattering, quantum evaporation, time-of-flight measurements of moving ions and mobility measurements of ions and electrons. This symposium emphasized the dynamic correlations in He II, which are characterized by correlation functions. The elementary excitations are defined as the poles of the single particle Green's function $G(k,\omega)$. By neutron inelastic scattering, the dynamic structure factor $S(k,\omega)$ is measured, which is the two-body correlation function of density fluctuations. Raman scattering measures the four-body correlation function of density fluctuations $H(k,k',\omega)$. Properties of these correlation functions and their mutual relations were discussed both theoretically and experimentally. Ab initio calculations of $S(k,\omega)$ were reviewed and a few models of temperature dependence of $S(k,\omega)$ were proposed. The coupling of $G(k,\omega)$ and $S(k,\omega)$ was pointed out as a key effect of Bose condensation on the dynamic properties of He II. Theories of the Raman scattering of He II were reviewed from various aspects. The process of scattering was clarified by the observation of the s-wave contribution. The fine structures in the Raman spectrum, confirmed in recent experiments, stimulated theoretical research into the relation between $S(k,\omega)$ and $H(k,k',\omega)$ as well as discussion of the relation between density fluctuations and quasiparticle operators. The importance of not only roton-roton but also roton-maxon and maxon-maxon interactions was stressed. In the following, short summaries of the papers are given.

Campbell and *Clements* review microscopic theories of strongly coupled Boson quantum fluids. They discuss that microscopic theories of the ground and low excited states and of dynamic structure factor $S(k,\omega)$ have been developed to the point of quantitative agreement with experiment. They report their recent work on the extension of the ab initio theory of the ground state to finite temperatures, where it provides a good account of the liquid-gas properties of the fluid. But a more sophisticated approach is needed to include

Springer Series in Solid-State Sciences, Vol. 79   **Elementary Excitations in Quantum Fluids**   1
Editors: K. Ohbayashi · M. Watabe   © Springer-Verlag Berlin, Heidelberg 1989

the λ transition in ^{4}He. They also discuss the theory of the second-order light scattering from the viewpoint of the microscopic theory.

Griffin discusses the nature of the elementary exitations in Bose fluid at finite temperatures, as well as their somewhat subtle relation to the density fluctuations, which can be studied by inelastic scattering probes. He stresses the importance of the Bose condensate in its role of coupling the resonance of $G(k,\omega)$ to the resonances of $S(k,\omega)$. He describes the role of the Bose condensation on the frequency and damping of the modes in both $G(k,\omega)$ and $S(k,\omega)$, and suggests that detailed experimental measurements of the temperature dependence of $S(k,\omega)$ would yield proof of his theory.

Ohbayashi reviews recent advances in the experimental study of Raman scattering from He II. The two-roton Raman scattering was measured at various pressures. Although the svp spectrum was confirmed to fit to the two-roton bound state model of Ruvalds-Zawadowski reasonably well, the spectra measured at higher pressures showed considerable deviations from the model. From the polarization properties, the spectra of s-wave scattering and d-wave scattering were determined. The integrated intensity ratio was found to agree with theoretical predictions by the Minnesota group. Fine structures were confirmed, which are interpreted as multiphonon processes. Comparisons with theories are given. The temperature dependence of the Raman spectrum is briefly described in relation to related behavior of $S(k,\omega)$. Future directions of the work are suggested.

Udagawa, *Watabe* and *Ohbayashi* report their recent experimental results on the polarization properties of liquid ^{3}He and ^{4}He. s-wave scattering was found in both of these fluids. The integrated intensity ratios of s-wave to d-wave agree with the theoretical predictions by the Minnesota group.

Svensson reported the temperature dependencs of $S(k,\omega)$ of He II. He compared his experimental results with the Woods-Svensson model, which is a two-fluid model of $S(k,\omega)$. The model was found to fit well with the experimental result at svp. However, at applied pressure of 20 bars, the fit was found to be less good.

Golub reports the upscattering of ultra-cold neutrons (UCN) by He II, which occurs on total reflection of UCN by material surfaces. The upscattering of UCN by He II is interesting because it allows the direct determination of the UCN density inside the helium and the study of the 3-photon interaction in the helium at k and T values inaccessible to other methods. The time dependence of the UCN density was observed to show a good agreement with the expression taking into account the upscattering rate.

Lauter, *Frank*, *Godfrin*, and *Leiderer* summarize their study on excitations in He films, including their recent results. Two layers adjacent to the graphite can be solidified due to the van der Waals attraction with it. Subsequent layers are found to be liquid, for which the excitations were found at energies below the bulk roton value. Because they are also visible in the neutron scattering of a thick film and even if the film is extended to bulk liquid, these excitations have to be localized near the solid-liquid interface. These modes show only little or no dispersion; they exhibit localized behavior. Only at a momentum transfer below the one of the bulk maxon has an excitation been detected which shows dispersion like an accoustic phonon with a reduced sound velocity with respect to bulk helium.

Halley discusses theories of Raman scattering. Historical development and present status are reviewed in relation to recent experimental results. He stresses that theoretical soundness of the two-roton bound state model is doubtful because it violates the excluded volume conditions. He discusses the relation of polarization properties and the microscopic inter-atomic interactions. His theory predicted the intensity ratio of s-wave to d-wave using the experimental width of the ultraviolet absorption line, which was verified by recent experiments. He assigns the fine structures in the Raman spectrum as multiphonon processes, and he points out two possible sources of the multiphonon Raman processes; anharmonicity in quasiparticles and the nonlinearity between density fluctuations and quasiparticle operators. He suggests a direction of future theories which should take into account both of these factors and be free from the violation of excluded volume conditions.

Iwamoto discusses the effect of the three-excitation states on the Raman scattering from He II. In order to interpret the fine structures in the Raman spectrum of He II, he proposed that those structures are due to multiphonon processes associated with two or more elementary excitations. He calculates the density of states and the effect of the interaction between quasiparticles, and derived possible forms of the anomalies associated with three-phonon Raman scattering. He predicts an infinite slope at energies $3\Delta_0, 2\Delta_0, \Delta_0 + 2\Delta_1$, and $3\Delta_1$, as well as $2\Delta_0$ and $2\Delta_1$ where Δ_0 and Δ_1 are respectively the energy of a single roton and a single maxon.

Ruvalds gives a brief review of the excitations in liquid ^{4}He, with emphasis on the interactions which were found to give two-roton resonances. The implications of this discovery were particularly relevant to the analysis of the Raman spectrum and neutron scattering experimental data. Later studies also demonstrated the inadequacy of the Born approximation as a result of the resonant structure, and hence extensive theoretical progress was achieved by various groups in understanding the transport properties associated with phonons and rotons. Attempts to derive a truly microscopic description of the roton coupling encountered difficulties. Nevertheless, semi-phenomenological inputs for the structure factor led to successful descriptions of the lowest order interactions.

Nagai theoretically discusses Pitaevskii's condition that the phonon-roton dispersion curve should terminate at a certain momentum Q_c. In actual neutron inelastic scattering experiments, however, the asymptotic branch was not observed up to the end point Q_c. He shows theoretically that, due to a finite linewidth of rotons, at finite temperature, Pitaevskii's branch ceases to exist at a temperature-dependent momentum $Q_c(T)$ which is smaller than Pitaevskii's Q_c.

Fukushima and *Iseki* discuss the experimental fact that the upper branch of $S(k, \omega)$ lies between $2\Delta_0$ and $2\Delta_1$ for $0.5\,\text{Å}^{-1} < k < 2.0\,\text{Å}^{-1}$. They identify this peak as a resonance of roton-maxon pairs. From the comparison, they derive the sign and strength of roton-maxon interactions.

Sridhar discusses the experimental fact that the liquid structure factor $S(k)$ started decreasing in height with an increase in its width as well, signalling a loss in spacial order, when the temperature is lowered below the lambda point T_λ. He theoretically treats $S(k)$ in the case of the Bogoliubov form single-particle excitation at finite temperature. The main structure factor peak is found to increase as the temperature is increased from below T_λ toward T_λ accompanied by a sharpening of the peak.

Ishikawa and *Yamada* discuss the two-component character of $S(k, \omega)$ in He II, which is called the Woods-Svensson model. Their theory starts from three basic equations of two-fluid dynamics. Their result of the separation of $S(k, \omega)$ into two terms, one proportional to the normal density and one proportional to the superfluid density, does not take the same form as that experimentally observed by Woods and Svensson.

Sasaki introduces "dressed-boson" operators as the unitary transformation of creation and annihilation operators of a helium atom assuming that the total Hamiltonian of liquid helium is completely diagonalized by unitary operator. He tries a new viewpoint of elementary excitations along the concept of the dressed-boson. He derives softening of the elementary excitation in the small momentum region as the temperature approaches T_λ.

Thellung compared the two formulations of hydrodynamic motion by Euler and Lagrange. Quantization leads to phonons, which in Euler's case carry a momentum and in Lagrange's case have no momentum. In spite of this difference, physically observable effects must be the same since the two representations are equivalent. He confirmed this conclusion in an explicit comparison of the two cases for the microscopic derivation of the two-fluid equations in liquid He II.

Wyatt developed the quantum evaporation method to examine phonons and rotons in superfluid ^{4}He. By the combination of quantum evaporation and time-of-flight measurements, he enabled the detection of long-lived high-energy phonons and rotons. They can propagate ballistically over distances $\sim 1\,\mathrm{cm}$ in liquid ^{4}He at low enough temperatures, allowing the study of the scattering of these excitations. The first study is the scattering of high energy phonons by thermal phonons whose energy is 2 orders of magnitude lower. This should lead to a quantitative understanding of the 4-phonon process, which hitherto has been inaccessible.

Williams discusses the role of vortex excitations in the critical region near the lambda transition of ^{4}He. A simple model using circular vortex rings has been constructed following the general ideas of Kosterlitz and Thouless, extended to the case of three dimensions. A real-space renormalization technique generates a power-law phase transition that satisfies the Josephson scaling relation and has the correct correlation length. The physical picture that emerges agrees with the original intuitive proposals of Feynman and Onsager, and with recent Monte Carlo simulations.

Hendry, *Lawson*, *Williams*, and *McClintock* discuss excitations created or grown by motion of an object in He II. In this case, the superfluidity of ^{4}He can in principle break down in two quite distinct ways. Either rotons may be produced above the Landau critical velocity or, alternatively, quantized vortices may be created or expanded. The creation process can be investigated very readily through studies of the motion of negative ions which, in view of their small size, are most unlikely to be affected by remanent vorticity, which are invariably contained in He II. They report and discuss an unexpected temperature dependence of the rate at which vortices are created by negative ions for pressures $P > 15\,\mathrm{bar}$.

Vinen proposes a new method to study ripplons, the quantized surface waves associated with a quantum fluid if it has a free surface. Less is known about the detailed properties and behavior of these excitations than about the other excitations in quantum fluids because of the lack of an appropriate experimental method of observation. He proposes to use ions and electrons trapped at the surface of superfluid helium as probes for the study of ripplons because the mobility of them is affected by ripplons.

Williams, *McClintock*, and *Hendry* show that a negative ion can exist in several different forms. The most reliable method of determining ion velocities and mobilities are based on time-of-flight measurements. From the measurements, an ion current vs time plot shows normal-ion, exotic-ion, and fast-ion signals. The structure of the normal negative ion is well established. However, the nature of the exotic ions and the fast ion are not well understood. There is also an interesting unsolved problem as to why these ions favor the creation of rotons in pairs instead of single separated rotons.

Theory of Liquid ^{4}He at Finite Temperatures

Microscopic Theory of Strongly Coupled Boson Quantum Fluids

C.E. Campbell and B.E. Clements

School of Physics and Astronomy, University of Minnesota,
116 Church St. S.E., Minneapolis, MN 55455, USA

1. INTRODUCTION

Quantum fluids such as the helium liquids, hydrogen liquids, nuclear matter and neutron matter are composed of particles which interact via a strong, short-ranged repulsion and a weaker long-ranged attraction. Consequently, a theory which seeks to produce a quantitative account of the properties of these fluids must of necessity deal with strong, short-ranged correlations induced by the interactions. Such a microscopic theory of the ground and low excited states (including multiple excitations) and of the dynamic structure factor $S(k,\omega)$ has been developed to the point of quantitative agreement with experiment, particularly for the boson fluids exemplified by liquid ^4He. This situation is reviewed briefly in section 2.

The theory of the second order light scattering is reviewed elsewhere in this volume by HALLEY. There it can be seen that recent experiments by OHBAYASHI *et al.* are calling for more attention to this important dynamic probe of fluids. We discuss the state of the microscopic theory briefly in section 3.

Section 4 contains a review of our recent work on the extension of the *ab initio* theory of the ground state to finite temperatures, where it is seen that the most obvious generalization provides a good account of the liquid-gas properties of the fluid (including the critical point), but is in need of a more sophisticated approach to include the λ transition in ^4He.

2. *AB INITIO* THEORY OF STRONGLY CORRELATED BOSON QUANTUM FLUIDS

We begin with a description of the system using a non-relativistic hamiltonian which contains two-body interactions:

$$H = \sum_{i=1}^{N} \frac{-\hbar^2}{2m} \nabla_i^2 + \sum_{i<j}^{N} V(r_{ij})$$

for an *N* particle system. It is well known that three-body interactions may contribute significantly in helium, but their inclusion is a straightforward matter which will not be discussed here in the interest of simplicity.

Denoting the ground state wave function by Ψ_0, as a general matter the excited states of the system with wave number k can be expressed in a Fourier series which can be reordered in terms of the density fluctuation operator ρ_k:

Springer Series in Solid-State Sciences, Vol. 79   **Elementary Excitations in Quantum Fluids**
Editors: K. Ohbayashi · M. Watabe   © Springer-Verlag Berlin, Heidelberg 1989

$$\Psi_k = F_k \Psi_0 = \left[A_k^{(1)} \rho_k + \sum_{\substack{q_1, q_2 \\ q_1 + q_2 = k}} A_{q_1, q_2}^{(2)} \rho_{q_1} \rho_{q_2} + \right.$$

$$\left. \sum_{\substack{q_1, q_2 q_3 \\ q_1 + q_2 + q_3 = k}} A_{q_1, q_2 q_3}^{(3)} \rho_{q_1} \rho_{q_2} \rho_{q_3} + \cdots + \sum_{\substack{q_1, \ldots, q_N \\ q_1 + \ldots + q_N = k}} A_{q_1, \ldots, q_N}^{(N)} \rho_{q_1} \cdots \rho_{q_N} \right] \Psi_0$$

where the density fluctuation operator is defined by

$$\rho_k = \sum_{j=1}^{N} e^{i k \cdot r_j} .$$

The coefficients $A_{q_1 \ldots q_n}^{(n)}$ are closely related to the Fourier coefficients of F_k, which is well defined for the boson fluid since Ψ_0 has no nodes.

Experimental information about the ground state is available in the form of the low temperature limit of the liquid structure function $S(k)$, obtainable from neutron and X-ray scattering experiments, and from polarized and depolarized second order light scattering (see papers by HALLEY; OHBAYASHI; IWAMOTO; RUVALDS; and by UDAGAWA *et. al.* herein). Neutron scattering experiments have also been designed to probe for the condensate fraction n_0. These topics are discussed further in the papers by SVENSSON; GRIFFIN; HALLEY; OHBAYASHI; IWAMOTO; RUVALDS; and UDAGAWA *et. al.* elsewhere in this volume. In recent years, the Green Function Monte Carlo (GFMC) simulations by KALOS *et al* [1] have provided effectively exact results for the liquid ^4He ground state (subject only to the numerical errors inherent in Monte Carlo simulations and the uncertainty about the true interaction between pairs or multiples of particles). These have been of enormous value in ascertaining the accuracy of proposed interactions and in assessing the validity of analytical theories such as the Jastrow-Feenberg theory discussed herein.

Neutron scattering measurements of the dynamic structure function, $S(k,\omega)$ (see SVENSSON herein) and the Raman spectrum $I_R(\omega)$ (see papers by HALLEY; OHBAYASHI; IWAMOTO; RUVALDS; and by UDAGAWA *et. al.* herein) are direct probes of the excitation spectrum, coupling most strongly through the density fluctuation operator ρ_k or pair density fluctuation operator $\rho_k \rho_q$ to the low lying excited states of the fluid.

Quantitatively accurate analytical theories of the ground state and low excited states of strongly coupled quantum fluids have all been based upon the Jastrow ansatz for the ground state:

$$\Psi_0 = \prod_{i<j}^{N} e^{\frac{1}{2} u(r_{ij})}$$

where the Jastrow pseudopotential $u(r_{ij})$ is determined variationally:

$$\frac{\delta}{\delta u(r)} \frac{\langle \Psi_0 | H | \Psi_0 \rangle}{\langle \Psi_0 | \Psi_0 \rangle} = 0 \; .$$

The original simple parametrization of $u(r_{ij})$ produced remarkably good agreement with the experimental binding energy, $S(k)$, and condensate fraction, although the small k behavior of $S(k)$ was incorrect [2]. When this variation is a full functional variation, producing an Euler-Lagrange equation for $u(r_{ij})$, the binding energy is moved to within 10% of the experimental result, while, much more significantly, $S(k)$ acquires the correct small k behavior [3]. Most importantly, the close relationship between the structure of the ground state and the elementary excitation spectrum emerges naturally. More about this below.

The obvious generalization of this Euler-Lagrange Jastrow approach to include higher order correlations in the form of three–body factors in the form,

$$\Psi_0 = \prod_{i<j}^{N} e^{\frac{1}{2}u_2(r_{ij})} \prod_{i<j<k}^{N} e^{\frac{1}{2}u_3(r_i, r_j, r_k)}$$

where u_2 and u_3 satisfy

$$\frac{\delta}{\delta u_2} \frac{\langle \Psi_0 | H | \Psi_0 \rangle}{\langle \Psi_0 | \Psi_0 \rangle} = 0 = \frac{\delta}{\delta u_3} \frac{\langle \Psi_0 | H | \Psi_0 \rangle}{\langle \Psi_0 | \Psi_0 \rangle}$$

has proven to account for most of the additional binding energy and improved the liquid structure function [3, 4]. KROTSCHECK [5] has recently done a careful analysis of this situation, confirming the previous conclusions with increased confidence. Once again, the close relationship between the structure of the ground state and that of the phonon-roton spectrum becomes evident [6, 7].

The earliest quantitative work on the phonon-roton excitation spectrum of ^4He invoked a trial function for excited states of the form

$$| k \rangle = \frac{\rho_k \Psi_0}{\sqrt{NS(k)}}$$

with the resulting Bijl-Feynman form of the spectrum:

$$\varepsilon_k^{BF} = \langle k | H - E_0 | k \rangle = \frac{\hbar^2 k^2}{2m\, S(k)}$$

where N is the number of particles and $S(k)$ is the ground state liquid structure function. As is well known, this result has the correct long-wavelength structure because of the linearity of the exact $S(k)$. However, this feature is missed if a trial function is used for Ψ_0 which does not behave properly at long wavelengths. It is a remarkable feature of the Jastrow function that, if it satisfies the Euler-Lagrange equation for u_2, then the expectation value of $H - E_0$ in state $/k>$ is exactly the same expression, except that the liquid structure function and E_0 which appear in this expression is that for the optimum Jastrow function [8]. Moreover, S(k) also has the proper long-wavelength behavior in this optimum Jastrow state, thereby giving ε_k^{BF} which has the proper small k limit.

10

Under the same conditions, the two Bijl-Feynman excitation state is given by

$$| k, q \rangle = \frac{\rho_k \rho_q \Psi_0}{\sqrt{N^2 S(k) S(q)}}$$

with excitation energy

$$\varepsilon_{k,q}^{BF} = \langle k, q | H - E_0 | k, q \rangle = \frac{\hbar^2 k^2}{2m\, S(k)} + \frac{\hbar^2 q^2}{2m\, S(q)} \; .$$

Of course the generalization of this produces a boson multi-excitation spectrum:

$$E = E_0 + \sum_k n_k \varepsilon_k \; .$$

The off-diagonal matrix elements in the corresponding states do not vanish, which is accounted for above in the expression $F_k \Psi_0$ for the exact phonon-roton state, including admixture of two and more BF states in the single excitation state. Indeed, the inclusion of bi-BF states produces a dramatic improvement in the agreement with the experimental spectrum, particularly when it is calculated with a trial ground state including the three-body pseudopotential u_3 discussed above [9]. This is apparently not an accident, since the Euler-Lagrange equation for the u_3 can be written in the simple form

$$\left\langle \Psi_0 \left| (H - E_0)\, \rho_k \rho_q \rho_{-k-q} \right| \Psi_0 \right\rangle = 0 ,$$

which is simply related to the vertex strength which couples the single BF state to the bi-BF state:

$$\left\langle \Psi_0 \left| \rho_k (H - E_0)\, \rho_q \rho_{-k-q} \right| \Psi_0 \right\rangle \; .$$

In particular, we found that the correct pressure dependence of the roton parameters (the gap, the curvature, and the location) is not obtained unless this Euler-Lagrange equation is satisfied [9]. This may be interpreted as an indication that the principal physical content of u_3 is the zero-point motion of the bi-BF contribution to the elementary excitation spectrum, a feature which is already well-understood at the Jastrow level with respect to the intermediate and long-range structure of u_2 [6, 7, 9].

The bi-excitation operator $F_{k,q}$ can of course be treated in an entirely analogous fashion, since it can be developed in a Fourier series similar to that of F_k:

$$F_{k\; p,\; p} = B_{k\; p,\; p}^{(0)} \rho_{k-p} \rho_p + B_k^{(1)} \rho_k + \sum_{\substack{q_1,\, q_2 \\ q_1 + q_2 = k}} B_{q_1,\, q_2}^{(2)} \rho_{q_1} \rho_{q_2} +$$

$$\sum_{\substack{q_1,\, q_2,\, q_3 \\ q_1 + q_2 + q_3 = k}} B_{q_1,\, q_2,\, q_3}^{(3)} \rho_{q_1} \rho_{q_2} \rho_{q_3} + \cdots + \sum_{\substack{q_1,\, \ldots,\, q_N \\ q_1 + \ldots + q_N = k}} B_{q_1,\, \ldots,\, q_N}^{(N)} \rho_{q_1} \cdots \rho_{q_N} \; .$$

MANOUSAKIS and PANDHARIPANDE [10] pointed out that the single excitation operator F_k can serve as a substitute for the ρ_k in $F_{k,q}$:

$$F_{k-p,\,p} = \overline{B}^{(0)}_{k-p,\,p}\, F_{k-p} F_p + \overline{B}^{(1)}_k\, F_k + \sum_{\substack{q_1,\,q_2 \\ q_1 + q_2 = k}} \overline{B}^{(2)}_{q_1,\,q_2}\, F_{q_1} F_{q_2} +$$

$$\sum_{\substack{q_1,\,q_2,\,q_3 \\ q_1 + q_2 + q_3 = k}} \overline{B}^{(3)}_{q_1,\,q_2,\,q_3}\, F_{q_1} F_{q_2} F_{q_3} + \ldots + \sum_{\substack{q_1,\,\ldots,\,q_N \\ q_1 + \ldots + q_N = k}} \overline{B}^{(N)}_{q_1,\,\ldots,\,q_N}\, F_{q_1} \ldots F_{q_N} \ .$$

The excited states of the system are observed through their effects at finite temperatures and are detected primarily through scattering probes such as neutron and photon scattering. With regard to neutron scattering, the cross section is proportional to the dynamic structure function $S(k,\omega)$, defined by

$$S(k, \omega) = \left\langle \rho_{-k}\, \delta(H - E_0 - \hbar\omega)\, \rho_k \right\rangle$$
$$= \frac{1}{\pi} \operatorname{Im} \left\langle \rho_{-k} \frac{1}{H - E_0 - \hbar\omega - i\eta} \rho_k \right\rangle$$
$$= \operatorname{Im} D(k, \omega) \ ,$$

where $D(k, \omega)$ is the density correlation response function. The calculation of D is facilitated by using a complete correlated basis, which is defined schematically by

$$|n\rangle = F_{k_1} F_{k_2} \ldots F_{k_n} \Psi_0 \ .$$

The earliest theory of $S(k,\omega)$ along this line was that of JACKSON [11], who used ρ_k for F_k and included basis contributions for $n \leq 2$. More recently, MANOUSAKIS and PANDHARIPANDE [12] used a correlated basis built from an F_k which includes $A^{(2)}$ in the Feynman-Cohen "backflow" form, with impressive results. Summarizing briefly, in an orthogonalized correlated basis representation the Hamiltonian is decomposed into a diagonal unperturbed part H_0 and the remaining off diagonal perturbation H_I. Standard perturbation theory may then be used to calculate $D(k,\omega)$. Including single and bi-excitations in their perturbation theory, they have

$$D(k,\omega) = D_{11}(k,\omega) + D_{12}(k,\omega) + D_{21}(k,\omega) + D_{22}(k,\omega)$$

where the subscript 1 indicates that the corresponding external point terminates in a single-excitation propagator and the subscript 2 indicates a bi-excitation propagator terminating at the external point. The result for $D_{11}(k,\omega)$, for example, is

$$D_{11}(k, \omega) = \frac{\left| \langle k | \rho_k | 0 \rangle \right|^2}{\varepsilon_F(k) + \Sigma_0(k, \omega) - \omega - i\eta}$$

where $\varepsilon_F(k)$ is the Feynman-Cohen spectrum, and

$$\Sigma_0(k, \omega) = -\frac{1}{2} \sum_{h,q} \left| \left\langle k | H - E_0 | h,q \right\rangle \right|^2 G_2^0(h,q,\omega)$$

where G^0_2 is the bi-excitation propagator. The bi-excitation contribution to the calculated $S(k,\omega)$ is at too high energy because the poles in G^0_2 are at the Feynman-Cohen energy. This is partially solved by including self-energy insertions in the bi-excitation propagator:

$$G_2(h,q,\omega) = \frac{1}{\varepsilon_F(h) + \Sigma_0(h, \omega - \varepsilon_F(h)) + \varepsilon_F(q) + \Sigma_0(q, \omega - \varepsilon_F(q)) - \omega - i\,\eta} \quad .$$

The phonon-roton spectrum obtained in this manner is in remarkably good agreement with the neutron scattering results of COWLEY and WOODS [13], although the quasiparticle strength $Z(k)$ still lies above the experimental value by a significant amount (approximately 20% near the roton wave-number). The results for $S(k,\omega)$ reported in the wave number range $0.8\ \text{Å}^{-1} < k < 2.0\ \text{Å}^{-1}$ show clear evidence of the bi-excitation structure. Most prominent in the lower end of this range are the two-maxon, two-roton, and maxon-roton peaks. At $k = 1.125\ \text{Å}^{-1}$ the maxon-roton maximum is the most prominent feature, with a small contribution evident from the two-maxon energy; there is no two-roton peak. However at $k = 1.525\ \text{Å}^{-1}$ the two-roton peak is comparable to the two-maxon peak, and there is lower frequency structure which is interpreted as roton-phonon and maxon-phonon. At $k = 1.925\ \text{Å}^{-1}$, the two-maxon and two-roton peaks are both evident , but the maxon-roton peak has disappeared entirely.

Throughout these calculations, the self-energy insertions move the structure to smaller ω and increase the intensity. However, the experiment seems to have still higher intensity, which might be due to the absence of four-excitation contributions to the propagator.

This calculation of $S(k,\omega)$ by MANOUSAKIS and PANDHARIPANDE [12] should be viewed as the state-of-the-art, which will not be qualitatively improved without enormous expenditures of computer time and effort to include three- and four-excitation contributions. It would be possible to repeat the calculation with a different choice of $A^{(2)}$ such as that used by CHANG and CAMPBELL [9]. However, the possible improvement doesn't merit the effort. More interesting would be to extend the calculation to $T > 0$, which may be possible within the variational density matrix theory described below.

3. SECOND ORDER LIGHT SCATTERING

Second order light scattering is a probe of both the statics and dynamics of fluids. The advantage of this probe is that it couples most directly to the bi-excitations of the fluid, and in the static limit is a direct probe of the three and four particle correlations in the fluid. However it suffers the well known disadvantage that the interaction of the probe with the fluid cannot be factored out of the signal, thus obscuring the response or correlation function which is characteristic of the fluid. This situation is exacerbated by the fact that the function which couples the light to the fluid--the pair polarizability tensor--is not known with much confidence. This places a higher responsibility on theory and simulations to interpret the experimental results. Although both the theoretical and experimental situation is discussed thoroughly elsewhere in this volume, we wish to make a few observations within the context of correlated theories.

For the convenience of the reader we adopt the notation that Halley uses in this volume, which differs in part from our publications.

The extinction coefficient for second order light scattering can be expressed in the form

$$h(\omega) \propto I_s(\omega) \, (\hat{\varepsilon}_0 \cdot \hat{\varepsilon}_n)^2 + I_d(\omega) \, [\frac{3}{4} + \frac{1}{4} (\hat{\varepsilon}_0 \cdot \hat{\varepsilon}_n)^2]$$

where $\hat{\varepsilon}_0$ and $\hat{\varepsilon}_n$ are the polarizations of the incoming and outgoing light, respectively, in the scattering process, and I_s and I_d are the two components of the extinction coefficient which can be separated by measuring the polarization $\hat{\varepsilon}_n$ of the scattered light:

$$I_\alpha(\omega) = c_\alpha \int 4\pi q^2 \, dq \int 4\pi q'^2 \, dq' \, t_\alpha(q) \, t_\alpha(q') \, G_\alpha(q, q'; \omega)$$

where $\alpha = s, d$; $c_s = 1$ and $c_d = {}^4/_{25}$; and G_α is the α Legendre projection of the fourier transform ($\mathcal{F.T.}$) of the pair-pair correlation function:

$$G(q, q'; \omega) = \mathcal{F.T.}\left[P_p(r, r'; t) \right]$$

where

$$P_p(r, r'; t) = \frac{1}{(N\rho^2)} \left[\left\langle \sum_{i \neq j}^{N} \delta(r - r_{ij}(0)) \sum_{m \neq n}^{N} \delta(r' - r_{mn}(t)) \right\rangle \right.$$
$$\left. - \left\langle \sum_{i \neq j}^{N} \delta(r - r_{ij}(0)) \right\rangle \left\langle \sum_{m \neq n}^{N} \delta(r' - r_{mn}(t)) \right\rangle \right]$$

where r_{ij} is the vector distance between particles i and j in the fluid. The functions $t_\alpha(q)$ are the fourier transforms of the two independent components of the pair-polarizability tensor, thus representing the coupling between the light and the fluid. Since these are not "contact" interactions, the extinction coefficients I_α are summed over all wave-numbers, making it impossible to directly measure G_α.

Information about correlations in the ground state is obtained from the total extinction coefficients:

$$I_\alpha = \int I_\alpha(\omega) \, d\omega$$

which are simply weighted integrals of the equal time pair-pair correlation functions:

$$I_\alpha = (2\pi)^6 c_\alpha \int r^2 \, dr \int r'^2 \, dr' \, t_\alpha(r) \, t_\alpha(r') \, P_{p,\alpha}(r, r'; t = 0)$$

where $P_{p,\alpha}$ is the α Legendre projection of P_p in terms of the angle between r and

Since I_α is a single number at each pressure and temperature, the useful information about fluid correlations is obtained only by examining the pressure and temperature dependence. Of more interest, however, is the ratio of polarized to depolarized intensity, I_s/I_d. There were early theoretical and experimental claims that $t_s(r)$ vanishes, so that $I_s(\omega)$ vanishes at at all frequencies. However O'BRIEN *et al* [14] and OXTOBY and GELBART [15] pointed out that the overlap of atomic wave functions will produce a significant polarized effect in the pair polarizability at small r (and at the same time modify the depolarized component at small *r*). KLEBAN and HALLEY [16] used data from ultraviolet absorption line shapes to extract a phenomenological estimate of I_s/I_d , and CAMPBELL *et al* [17] used the earlier calculation of the pair-pair correlation function for the ^4He Jastrow wave function to actually calculate the ratio I_s/I_d using the t_s and t_d available from references [14] and [15]. The prediction of CAMPBELL *et al.*[17]

$$0.08 \leq I_s/I_d \leq 1.88$$

encompasses the experimental result of 0.10 subsequently obtained by the Hiroshima group [18]. The wide range in the theoretical estimate is entirely due to the uncertainties in the pair-polarizability tensor components t_s and t_d.

It should be pointed out that the "classical" dipole-induced-dipole part of $t_d(r)$ behaves like r^{-3}, while the quantum mechanical contribution due to the overlap of closed atomic shells are enormously large at short distances, and are often represented by a divergent term going as r^{-6}. Thus the functions $t_\alpha(r)$ are not fourier transformable, and calculations of the extinction coefficients should actually be carried out in coordinate space as we have described above, where the short-range correlations in $P_p(r, r')$ naturally compensate this effect. Of course these short range correlations are induced by the two-body interaction, which is large at small r for exactly the same reasons as with $t_\alpha(r)$. Consequently, it is not necessary to pay extra attention to the "excluded volume condition" (see HALLEY in this volume) if the work is done in coordinate space and proper short-range correlations are included in $P_p(r, r')$.

Nevertheless, the physical interpretation of the dynamic second-order light scattering is much more natural in momentum space, where the fourier transform of the pair-pair correlation function is seen to be the correlation function for pairs of density fluctuation operators with equal and opposite momentum:

$$G(\mathbf{q},\mathbf{k};t) = \frac{1}{(N\rho)^2}\left[\left\langle \rho_\mathbf{q}(0)\rho_{-\mathbf{q}}(0)\,\rho_\mathbf{k}(t)\rho_{-\mathbf{k}}(t)\right\rangle - \left\langle \rho_\mathbf{q}(0)\rho_{-\mathbf{q}}(0)\right\rangle\left\langle \rho_\mathbf{k}(t)\rho_{-\mathbf{k}}(t)\right\rangle\right].$$

As discussed in section 2, the frequency transform of this function should show a strong resonance at the bi-excitations (e.g. two-roton) energies, similar in some regards to the bi-excitation density of states, though with substantial weighting (and possibly ckowing) duo to tho oorrolationo.

A cumulant analysis of G leads to the following structure [19]:

$$(N\rho)^2 G(\mathbf{q},\mathbf{k};t) = \left|\left\langle \rho_\mathbf{k}(0)\rho_{-\mathbf{k}}(t)\right\rangle\right|^2 [\delta_{\mathbf{k},\mathbf{q}} + \delta_{\mathbf{k},-\mathbf{q}}] + N F_p(\mathbf{k},-\mathbf{k};\mathbf{q},-\mathbf{q};t)$$

where F_p includes fundamental correlations between the two density fluctuation pairs. This suggests as a first approximation that F_p be omitted, thereby treating the bi-excitations as if they are not correlated. In that case, the frequency transform of the first two terms in G would be simply the convolution of two dynamic structure functions:

$$\rho^2 \, \widetilde{G}(\mathbf{q},\mathbf{k};\omega) \approx [\,\delta_{\mathbf{k},\mathbf{q}} + \delta_{\mathbf{k},-\mathbf{q}}\,] \int S(\mathbf{k},\omega-\omega')\, S(\mathbf{k},\omega')\frac{d\omega'}{2\pi} \ .$$

In this approximation, the frequency structure of G would contain no additional information beyond that contained in $S(k,\omega)$, and would primarily reflect the convolution of two single resonance poles, one from each $S(k,\omega)$.

STEPHEN [20] used this approximation to calculate $I_d(\omega)$, except that he introduced a cutoff in $t_d(r)$, setting it to zero for r less than the hard core radius of the helium interaction. It is not at all clear, however, that the approximation takes proper account of the correlations between the excitation pairs, and will certainly omit any information about possible bound roton pairs which is not already present in $S(k,\omega)$.

To investigate the importance of F_p, PINSKI and CAMPBELL calculated it in the static limit $(t = 0)$ using optimized Jastrow functions for the ground state as a function of density. There it was found that the s and d components of $F_p(\mathbf{k},\mathbf{q})$ (called $U_4(\mathbf{k},\mathbf{q})/N$ in that work) have a large and negative extremum at $k = q \approx 2\ \mathring{A}^{-1}$, i.e., in the vicinity of the roton wave number.

While this is very suggestive vis a vis roton interactions, a full time or frequency dependent calculation, similar to that for $S(k,\omega)$, needs to be carried out within the correlated framework for liquid ^4He. While simulations are entirely feasible for classical fluids using molecular dynamics (a simulation for liquid argon is nearing completion [21]), a successful quantum molecular dynamics simulation method for a fluid has apparently not yet been developed; the GFMC[1] and path integral Monte Carlo simulations [22] have not been applied to dynamical simulations at any level, nor have they been used to simulate the static pair-pair correlation functions or total extinction coefficients for ^4He.

The most promising microscopic analytical theory would be to adapt the correlated basis function approach of MANOUSAKIS and PANDHARIPANDE [12] to calculate the pair-pair dynamic structure function:

$$(N\rho)^2 \, \widetilde{G}(\mathbf{q},\mathbf{k};\omega) = \Big\langle \rho_{\mathbf{q}}\rho_{-\mathbf{q}}\, \delta(H - E_0 - \hbar\omega)\, \rho_{\mathbf{k}}\rho_{-\mathbf{k}} \Big\rangle$$

$$= \frac{1}{\pi}\, \mathrm{Im} \Big\langle \rho_{\mathbf{q}}\rho_{-\mathbf{q}} \frac{1}{H - E_0 - \hbar\omega - i\eta}\, \rho_{\mathbf{k}}\rho_{-\mathbf{k}} \Big\rangle$$

$$= \mathrm{Im}\, D_p(\mathbf{q},\mathbf{k};\omega) \ .$$

As discussed above, the leading term will be the frequency convolution of two dynamic correlation functions $S(k,\omega)$ which only contributes for $\mathbf{k} = \pm\,\mathbf{q}$, while the fundamental correlations will be obtained from the frequency transform of $F_p(\mathbf{k},\mathbf{q};t)$:

$$S_p(\mathbf{q},\mathbf{k};\omega) = \frac{1}{2\pi} \int e^{-i\omega t}\, F_p(\mathbf{k},\mathbf{q};t)\, dt$$

$$= (N\rho)^2 \, \widetilde{G}(\mathbf{q},\mathbf{k};\omega), \quad \mathbf{k} \neq \pm\,\mathbf{q} \ .$$

A somewhat simpler analysis of the frequency dependence can be obtained from experimental data or the static calculations by calculation of the frequency moments [23]:

$$I_\alpha^{(n)} = \int \omega^n \, I_\alpha(\omega) \, d\omega \, .$$

In particular, the f-sum rule is obtained from the double commutator with H in the usual manner:

$$(N\rho)^2 \int \omega \widetilde{G}(\mathbf{q},\mathbf{k};\omega) \, d\omega = \left\langle \left[\rho_q \rho_{-q}, [H, \rho_k \rho_{-k}]\right] \right\rangle$$

$$= \frac{-\hbar^2 \mathbf{k}\cdot\mathbf{q}}{m} \left\{ \left\langle \rho_q \, \rho_k \rho_{-k-q} \right\rangle - \left\langle \rho_q \rho_{-k} \, \rho_{k-q} \right\rangle - \left\langle \rho_{-q} \, \rho_k \rho_{q-k} \right\rangle + \left\langle \rho_{-q} \, \rho_{-k} \, \rho_{q+k} \right\rangle \right\}$$

which also must be analyzed differently if $\mathbf{k} = \pm\, \mathbf{q}$, in which case the contribution to the sum rule comes directly from the f sum rule for $S(k, \omega)$:

$$\rho^2 \int \omega \widetilde{G}(\mathbf{q},\mathbf{k};\omega) \, d\omega = \frac{2\hbar^2 k^2}{m} S(k) + O\left(\frac{1}{N}\right) \, .$$

(Although it might appear that this is the only important term since it is higher order in N than the three-ρ terms, these latter terms contribute to the leading order in N when integrated over $\mathbf{k}$ and $\mathbf{q}$, and it is these terms which carry the information about the correlations between the quasiparticles.)

Again we emphasize that, as a practical matter, the short range structure of the $t_\alpha(r)$ make it necessary to carry out the theoretical calculation of this moment in coordinate space. The fourier transform of the three-ρ functions will introduce the well-known three-body distribution function, $\rho_3(\mathbf{r}_1, \mathbf{r}_2, \mathbf{r}_3)$ as well as the radial distribution function $g(r)$, both of which behave sufficiently well at small interparticle distances to compensate the divergent behavior of t_α.

4. ^{4}He AT FINITE TEMPERATURES

The theory described above has all been developed for a Bose fluid at $T = 0$. A method for extending this theory to finite temperatures has been developed during the past few years, but has far to go to reach the depth of the the $T = 0$ theory.

The objectives of such a theory are fairly clear: an *ab initio* accounting of the phase diagram of the fluid, together with the temperature dependence of the measurable quantities such as the excitation spectrum, dynamic structure function, liquid structure function, extinction coefficient for second order light scattering, and condensate fraction. Interesting features in the experimental data which have drawn attention are the sharpening of the first maximum in the liquid structure function as the temperature increases below the λ temperature (indicating an increase in the positional ordering, in contrast with ordinary fluids) and the substantial decrease of the roton gap as T approaches T_λ from below.

In work which began several years ago as a collaboration with Kurten, Senger and Ristig (CAMPBELL *et al*, [24]) we have developed a variational statistical mechanical approach which is the generalization of the Jastrow-Feenberg Euler-

Lagrange approach described above which has been so successful for the ground state. The most important result of this work to date is that it is the first microscopic calculation which accurately determines the liquid-gas critical point and spinodal line for a real quantum fluid. In parallel, CEPERLEY and POLLOCK [22] have developed a powerful Green function path integral method to simulate liquid ^4He at finite temperature. In this paper we will briefly review our theoretical results, which are not yet sufficiently developed to make direct comparison to the simulations [24, 25].

We began with the general observation that the density matrix in coordinate space representation can be written as

$$W(r_1...r_N \, ; \, r'_1...r'_N) = \frac{\Phi^*(r_1...r_N) \, Q(r_1...r_N \, ; \, r'_1...r'_N) \, \Phi(r'_1...r'_N)}{Z}$$

where the incoherence function Q is a function which cannot be factored into functions of the unprimed coordinates times functions of the primed coordinates. In the boson case, each of the factors in W is real and positive, and is symmetric individually in the primed coordinates and the unprimed coordinates, which fits nicely with a Jastrow type of analysis.

The most powerful variational principle in this context is the Gibbs-Delbruck-Moliere minimum principle, which states that the trial Helmholtz free energy, F_t is bounded below by the true free energy F_0:

$$F_0 \leq F_t = \mathrm{tr}\,(H\,W_t) + \frac{1}{\beta}\,\mathrm{tr}\,(W_t\,\mathit{ln}\,W_t)$$

where $\beta = (k_B T)^{-1}$ and W_t is a suitable trial density matrix. The second term in F_0 is $-S_t/T$, where S_t is the exact entropy of the trial density matrix. It is the calculation of this entropy which presents the most difficulty in our work.

The formal expression of the Euler-Lagrange equations for the density matrix is

$$\frac{\delta F_t}{\delta \Phi} = 0 \quad , \quad \frac{\delta F_t}{\delta Q} = 0 \quad .$$

If these equations could be solved without restriction, the result would be the exact density matrix of the system.

The minimum choice of trial density matrix which can handle a strong, short-ranged interaction such as is present in liquid ^4He is to choose a product of two-body functions for both Φ and Q, giving

$$W_t\,(r_1...r_N \, ; \, r'_1...r'_N) = \frac{1}{Z_t} \prod_{i<j}^{N} e^{\frac{1}{2}u(r_i - r_j)} \prod_{i,j}^{N} e^{\omega(r'_i - r_j)} \prod_{i<j}^{N} e^{\frac{1}{2}u(r'_i - r'_j)}$$

The functions $u(r)$ and $\omega(r)$ are determined by the coupled set of Euler-Lagrange equations for the free energy:

$$\delta F_t / \delta u(r) = 0 = \delta F_t / \delta \omega(r) \, .$$

Consequently, the solutions $u(r)$ and $\omega(r)$ of these equations are temperature dependent. At zero temperature $\omega(r)$ vanishes and $u(r)$ becomes the ground state Jastrow pseudopotential discussed in section 2 above.

This two-body structure of Q was first derived by PENROSE [26], and later by REATTO and CHESTER [27], by calculating the contributions of phonons to the density matrix. It is reassuring to note that our solution of the Euler Lagrange equations reproduces their results in the low temperature limit.

Expressing the entropy in terms of $u(r)$ and $\omega(r)$ is accomplished by using the replica method:

$$S_t = -k_B \, \mathrm{tr} \left(W_t \, \ell n \, W_t \right) = -k_B \frac{d}{d\sigma} \, \mathrm{tr} \, W_t^\sigma \Big|_{\sigma = 1}$$

where

$$\mathrm{tr} \, W_t^\sigma = \frac{1}{Z_t^\sigma} \int \prod_{n=1}^{\sigma} \left[d(\, r_{1_n} .. r_{N_n}) \prod_{i_n < j_n}^{N} e^{u(r_{i_n} - r_{j_n})} \prod_{i_n , j_{n+1}}^{N} e^{\omega(r_{i_n} - r_{j_{n+1}})} \right]$$

and $r_{j_{n+\sigma}} = r_{j_n}$ for $n \le \sigma$.

Identifying the trace above as the configuration integral for a *classical* σ component fluid with effective interactions $u(r)$ and $\omega(r)$ enables us to express the $u(r)$ and $\omega(r)$ functional derivatives of the entropy in terms of σ derivatives of the multicomponent two-body distribution functions.

The implementation of this procedure for calculating the entropy is greatly facilitated by an approximation which amounts to the decoupling of the fourier components of the entropy fluctuations. The result is a Bose fluid entropy:

$$S_t = \sum_k (1 + n_k) \, \ell n \, (1 + n_k) - n_k \, \ell n \, n_k$$

where the Bose occupation number is related to $u(r)$ and $\omega(r)$ by

$$n_k (n_k + 1) = \tilde{\omega}(k) \, S(k)$$

where $\tilde{\omega}(k)$ is the fourier transform of $\omega(r)$. Introducing a statistical, temperature dependent excitation spectrum $\varepsilon(k)$ defined by

$$n_k = \frac{1}{e^{\beta \varepsilon(k)} - 1} \quad ,$$

the Euler-Lagrange equations give an equation for $\varepsilon(k)$:

$$\varepsilon(k) = \frac{h^2 k^2}{2m \, S(k)} \coth \frac{\beta \varepsilon(k)}{2}$$

where $S(k)$ is the temperature dependent liquid structure function. (A completely independent derivation of this result is obtained by making the single-resonance ansatz for the density-density correlation function along with the requirement that the f-sum rule be satisfied.) This result is the finite temperature generalization of the Bijl-Feynman expression given above. As should be the case for finite temperature in the small k limit, $S(k) \rightarrow k_B T/mc^2$, and thus $\varepsilon(k) \rightarrow \hbar ck$ as is observed experimentally.

The solution of the Euler-Lagrange equations within this approximation produces a liquid-gas phase diagram with a well-defined spinodal line and critical point. The spinodal line is defined by the locus of points for which the coefficient c (the isothermal sound velocity) of k in $\varepsilon(k)$ vanishes. The interior of the region bounded by the spinodal line consists of thermodynamic points for which the Euler-Lagrange equations have no solutions.

The existence of the spinodal line is already implied by the spinodal point at $T = 0$ which is obtained by the ground state analysis, together with the existence of the liquid-gas critical point, which also lies on the spinodal line. Of course the ground state analysis does not involve the approximation for the entropy, and thus an improved approximation will still terminate at $T = 0$ at the same spinodal point.

The approximation used in the entropy calculation should yield reliable results when the number of excitations are few relative to the particle number. Thus one might expect it to fail at low density for a fixed temperature, or at high temperature for a fixed density. The shortcomings which result from using this approximation are clearly evident in the results reported in [25], where it is observed that the spinodal line for ^4He extrapolates to a finite temperature at zero density. A primary consequence of our present work on an improved approximation scheme [28] is that our spinodal line agrees more closely with expectations in the low temperature and density regime of the thermodynamic phase diagram.

Briefly summarizing our present efforts, the two Euler-Lagrange equations for the density matrix involve $d\, S_{\alpha\beta}(k; \sigma)/d\sigma\,|_{\sigma = 1}$, for $\alpha,\beta = 1,1$ and $1,2$, because of the entropy term. The chief requirement is that expressions be developed for $S_{\alpha\beta}(k; \sigma)$, which can be analytically continued to non-integer σ. Using a diagrammatical analysis, it is a straight-forward task to show that the diagrams which comprise the approximation described above are a subset of the diagrams that form multicomponent hypernetted chain equations (HNC). A lengthy analysis leads to a reordering of the HNC equations which exhibit this earlier result as the first term in equations which are amenable to improvement via an iterative scheme. With regard to the excitation spectrum, the Bijl-Feynman form becomes

$$\varepsilon_k = \frac{\hbar^2 k^2 \coth(\beta\,\varepsilon_k/2)}{2m\,S(k)\,[\,1 + Q(k)\,]} \quad .$$

The correction is incorporated in the $Q(k)$ term in the factor in the denominator, which produces a small improvement in the spectrum. However, the most successful aspect of this result is that the spinodal line is much improved in the low density regime.

It is now clear that, without much additional effort, the optimum Penrose-Reatto-Chester-Jastrow (PRCJ) density matrix for liquid ^4He may be calculated along with the associated $S(k)$ and $\varepsilon(k)$ and condensate fraction. A final calculation using Monte Carlo techniques would remove all approximations within this density matrix function space, and is well within current computational capabilities. The nature of the results of such a calculation are already evident: the liquid-gas phase diagram, including the critical point, will be well described; the condensate fraction will have the correct low temperature behavior; the phonon-roton curve will be qualitatively correct--quantitatively correct at long wavelengths--but in error near the roton minimum by a factor of two; and the liquid structure function will sharpen with increasing temperature, though not as much as seen in experiment. Introducing three-body factors in Φ and "two plus one"-body factors in the incoherence factor Q will quantitatively improve all of these experimental results in much the same way as the ground state and phonon-roton spectrum were improved in the $T = 0$ calculations above. Indeed, BATTAINI and REATTO [29] have already examined the effect on S(k) of including three- and four-body factors generated by the thermal occupation of Feynman-Cohen excitations in the N-body probability density, with the same conclusions.

There is a more serious problem with this class of trial density matrices which must be answered before much further effort is justified. There is no evidence of a λ-type transition or a Bose-Einstein transition in any of the calculations. Indeed, it is straightforward to show that there is always a Bose-Einstein condensate present in these density matrices. The fault is most clearly exhibited by recalling that the high temperature limit of a Bose fluid density matrix manifests the absence of off-diagonal-long-range-order (ODLRO) through the presence of a permanent of short ranged functions in the incoherence factor Q :

$$Q(\{r\}; \{r'\}) = \text{Perm } \Gamma(| \, r_i - r_j' \, |)$$

where $\Gamma(r)$ is a Gaussian with length scale given by the thermal deBroglie wavelength. This structure is not well approximated by the product of pair functions which defines Q in a PRCJ density matrix. The important feature of this incoherence factor which kills the ODLRO is the fact that Γ decreases to zero with increasing r. If the asymptotic value of Γ were finite, the density would have a Bose-Einstein condensate and ODLRO. Indeed, the value of the condensate would be proportional to this asymptotic value. In his contribution to these proceedings, GRIFFIN points out that the role of the condensate in Jastrow-Feenberg theories is obscure at best, and thus they may not be suitable for elucidating the physics of the ODLRO phase transition. However, proceeding along the lines just described may overcome this problem, since it is clear that the very structure of the incoherence factor is directly related to the presence/absence of a condensate. This class of trial density matrices should provide the first step toward a microscopic theory of ^4He which includes its most interesting feature, the λ transition. Unfortunately, the mathematical difficulties which must be faced are more difficult than with the PRCJ density matrix, and bear a certain resemblance to the challenges faced in the fermion liquid ^3He ground state theory.

ACKNOWLEDGEMENTS

One of us (CEC) would like to thank Professors Ohbayashi and Watabe for organizing this stimulating symposium, and making it possible for him to participate. This research was supported in part by the National Science Foundation and the Minnesota Supercomputer Institute.

REFERENCES

1. M. H. Kalos, Nucl. Phys. A328, 153 (1979) ; P. A. Whitlock, D. M. Ceperley, G. V. Chester and M. H. Kalos, Phys Rev. B19, 5598 (1979).
2. W. L. McMillan, Phys. Rev. A 138, 442 (1965)
3. C. C. Chang and C. E. Campbell, Phys Rev. B 15, 4238 (1977) and references cited therein.
4. C. E. Campbell, Phys. Lett. A 44, 471 (1973).
5. E. Krotscheck, Phys. Rev. B 33, 3158 (1986).
6. L. Reatto, Nucl. Phys. A 328, 253 (1979).
7. C. E. Campbell and F. J. Pinski, Nucl. Phys. A 328, 210 (1979).
8. C. E. Campbell and E. Feenberg, Phys Rev. 188, 396 (1969).
9. C. C. Chang and C. E. Campbell, Phys Rev. B 13, 3779 (1976).
10. E. Manousakis and V. R. Pandharipande, Phys. Rev. B 30, 5062 (1984).
11. H. W. Jackson, Phys. Rev. A 8, 1529 (1973).
12. E. Manousakis and V. R. Pandharipande, Phys. Rev. B 33. 150 (1986).
13. R. A. Cowley and A. D. B. Woods, Can J. Phys 49, 177(1971).
14. E. F. O' Brien, V. P. Gutschick, V. McKoy and J. P. Mctague, Phys. Rev. A8, 690 (1973).
15. D. W. Oxtoby and W. M. Gelbart, Mol. Phys. 30, 535 (1975).
16. P. Kleban and J. W. Halley, Phys. Rev. B 11, 3520 (1975).
17. C. E. Campbell, J. W. Halley, and F. J. Pinski, Phys. Rev. B 21, 1323 (1980).
18. M. Udagawa, H. Nakamura, M. Murakami, and K. Ohbayashi, Phys. Rev. B34, 1563 (1986).
19. F.J. Pinski and C. E. Campbell, J. Phys. (Paris) 39, C6-233 (1978).
20. M. J. Stephen, Phys. Rev. 187, 279 (1969).
21. F.J. Pinski and C. E. Campbell, Phys Rev. A 33. 4232, (1986); B. E. Clements, C. E. Campbell, P. Samsel and F. J. Pinski, work in progress.
22. D. M. Ceperley and E. L. Pollock, Phys. Rev. Lett. 56, 351 (1986).
23. C. E. Campbell and J. W. Halley, work in progress.
24. C. E. Campbell, K. E. Kurten, M. L. Ristig, and G. Senger, Phys. Rev. B 30, 3728 (1984).
25. G. Senger, M. L. Ristig, K. E. Kurten, and C. E. Campbell, Phys Rev. B 33 7562 (1986).
26. O. Penrose: In Proceedings of the International Conference on Low Temperature Physics, ed. by J. R. Dillinger (University of Wisconsin Press, Madison, Wisconsin, 1958), p. 117.
27. L. Reatto and G. V.Chester, Phys. Rev. 155, 88 (1967).
28. B. E. Clements and C. E. Campbell, work in progress.
29. S. Battaini and L. Reatto, Phys. Rev. B28, 1263 (1983).

Elementary Excitations in Bose-Condensed Liquids and Gases at Finite Temperatures

A. Griffin

Department of Physics, Kyoto University, Sakyo-ku, Kyoto 606, Japan
Department of Physics, University of Toronto, Toronto, Ontario, Canada*

1. INTRODUCTION

The major goal of this article is to review what the field-theoretic analysis of Bose-condensed fluids (liquids and gases) tells us about the nature of the excitations, both above and below the superfluid transition temperature. At $T = 0$, this analysis was initiated by Bogoliubov in 1947 for a weakly interacting dilute Bose gas (WIDBG) and the resulting picture is discussed in all advanced textbooks on statistical mechanics. Much less well known are the dynamical properties of a WIDBG at intermediate temperatures (where the condensate fraction may be strongly depleted). While the field-theoretic analysis was formally extended by Beliaev in 1957 to deal with Bose liquids, it was only in the early seventies that one began to understand the somewhat subtle (but absolutely crucial) role that the Bose broken symmetry has on the nature of the elementary excitations in superfluid ^{4}He. I have given a review [1] of certain aspects of resulting scenario at the Banff International Conference on Quantum Fluids and Solids. The present article may be viewed as part two of this account, with special emphasis on the nature of the elementary excitations in Bose fluids at finite temperatures, as well as their somewhat subtle relation to the density fluctuations [resonances in $S(\vec{Q},\omega)$] which can be studied by inelastic scattering probes. I refer to my Banff article for more detailed references to the technical literature. I minimize the use of equations but rather try to describe in physical terms what the microscopic theory implies.

The suggestion that the Bose broken symmetry [the average value of the field operator $< \hat{\psi}(\vec{r}) >$ being finite] is crucial to understanding the nature of the elementary excitations in superfluid ^{4}He is bound to find a certain resistance among people who study liquid ^{4}He. Without too much exaggeration, one can say that most experimental work on superfluid ^{4}He seems quite consistent with what one might call the "Landau Scenario". By this, I mean that superfluid ^{4}He is viewed as a gas of weakly interacting phonons and rotons, whose dispersion relation is given directly by neutron scattering. As we all know, at low temperatures (say, $T \lesssim 1.7K$), this theoretical picture has been very successful in understanding both the equilibrium and transport

---

* permanent address

Springer Series in Solid-State Sciences, Vol. 79   **Elementary Excitations in Quantum Fluids**
Editors: K. Ohbayashi · M. Watabe   © Springer-Verlag Berlin, Heidelberg 1989

properties of superfluid ^{4}He. It seems difficult to understand any criticism of this quasiparticle picture. It is only when one goes to higher temperatures and tries to understand the differences between the elementary excitations in liquid ^{4}He above and below T_λ (which must be different since, for example, liquid ^{4}He is a superfluid at $2K$ but non-superfluid at $2.3K$) that one is forced to develop a more complete understanding of the Landau quasiparticle picture that has served so well at lower temperatures. The same sort of conclusion is discussed in this Symposium by CAMPBELL in his review of recent attempts to understand the excited states at finite temperatures within the CBF approach.

Evidence that superfluid ^{4}He has a condensate has been slowly accumulated over the last 30 years [1]. Using inelastic neutron scattering at high momentum transfers (where the impulse approximation is valid), the magnitude and temperature dependence of the condensate fraction $n_0(T)$ can be extracted from the momentum distribution of ^{4}He atoms. This data [2] shows clearly that n_0 is very small (zero) above T_λ and builds up to a value (as T is lowered) which is roughly consistent with the best computer simulation values, namely $n_0(T = 0) \simeq 9\%$. Unfortunately much of the literature on this "condensate saga" discusses the condensate in isolation, as just "another" property of superfluid ^{4}He, rather than as evidence for Bose broken symmetry and all that this implies.

A major development in the last year involves Monte Carlo studies at finite temperatures (using Feynman path integral techniques) carried out by CEPERLEY and POLLOCK [3,4]. Their results include:

(a) The specific heat has the characteristic λ-singularity close to $T_\lambda = 2.17K$.

(b) The condensate fraction $n_0(T)$ is essentially zero above T_λ and then rapidly builds up below T_λ to a value consistent with 9% at $T = 0$.

(c) The normal fluid density $\rho_N(T)$ is in excellent agreement with experimental results. In particular, $\rho_N(T) \to 0$ as $T \to 0K$.

The small sample size (64 atoms) leads to some characteristic finite-size rounding near T_λ but even so, these preliminary results must be viewed as a major watershed in our understanding of superfluid ^{4}He. As I have discussed in my Banff paper [1] in more detail, the successful ab-initio calculation of the normal and superfluid densities is especially important. POLLOCK and CEPERLEY [4] essentially started with an exact relation which expresses $\rho_S(T)$ as the difference between the long wavelength, zero frequency momentum current response functions to transverse and longitudinal perturbations. This fundamental definition of $\rho_S(T) \equiv \rho - \rho_N(T)$ has been understood since the early sixties [5], in particular the fact that $\rho_S \neq 0$ is a direct consequence of a Bose broken symmetry $< \hat{\psi}(\vec{r}) > \neq 0$. This definition of $\rho_S(T)$ is not dependent on a simple quasiparticle picture being valid (as it isn't near T_λ), although at low temperatures it can be shown to reduce to Landau's well known formula in terms of excitations.

In my opinion, this work of CEPERLEY and POLLOCK [3,4] gives the most direct evidence that we have so far that the superfluid transition in ^{4}He is indeed associated with the formation of a Bose broken symmetry. As such, it forces us to look at superfluid ^{4}He from an aspect which has been largely untouched up to now - namely, to understand the changing properties of superfluid ^{4}He as the temperature increases as the <u>direct</u> consequence of the condensate decreasing in magnitude, vanishing at T_λ. In taking up this challenge, one must understand what the field-theoretic analysis of a Bose-condensed liquid says about the elementary excitations and we first review this in Section 2. We mention several suggestions for future experiments to verify various aspects of this theory. In Section 3, we sketch what the same formalism predicts for a WIDBG at finite temperatures [6]. Quite apart from the possible physical realization of such Bose gases in the laboratory over the next few years, the contrast between the nature of the elementary excitations in Bose gases and liquids gives one considerable insight into the role of the Bose condensate.

2. EXCITATIONS IN A BOSE LIQUID

To begin with, we recall that in the field-theoretic analysis, the elementary excitations are defined in a very precise manner as the resonances (poles) of the single-particle Green's function $G(\vec{Q},\omega)$. In a Bose-condensed system, this is a 2×2 matrix to deal with the anomalous field correlation functions associated with the Bose broken symmetry. In contrast, the density fluctuations are resonances of the density correlation function $S(\vec{Q},\omega)$, which is a two-particle Green's function. While $S(\vec{Q},\omega)$ can be probed by inelastic scattering techniques, the more basic field fluctuation spectrum described by $G(\vec{Q},\omega)$ cannot be probed so directly.

Referring to the approach discussed by CAMPBELL in this Symposium, we immediately see the difficulty of bringing these two approaches into contact. The CBF approach may be viewed as the modern development of the variational approach first used by Feynman. It works directly with either density fluctuations, or with variables related to them, in trying to find good approximations for the excited states. However the most natural way of dealing with the effect of Bose condensation is to work directly with the field operators, which are described by the single-particle Green's function $G(\vec{Q},\omega)$. Higher order correlation functions like $S(\vec{Q},\omega)$ are then expressed in terms of the more basic correlation function $G(\vec{Q},\omega)$. It would seem that, somehow or other, the CBF approach must be generalized to deal with the field fluctuation spectrum more directly. It would also be interesting if such methods were used to calculate the superfluid density (as POLLOCK and CEPERLEY [4] have done) to verify directly that the CBF excited states give the correct value of ρ_S. As we mentioned in Section 1, the superfluid density is in some ways a more direct consequence of the existence of the Bose broken symmetry than the condensate fraction is.

We now summarize, in a qualitative way, the changes introduced in the excitation spectrum of $G(\vec{Q},\omega)$ and $S(\vec{Q},\omega)$ when we pass though T_λ (See also Sections 2 and 3 of GRIFFIN [1]). For $T > T_\lambda$, $G(\vec{Q},\omega)$ has a broad, possibly particle-like, spectrum whose characteristic frequency we denote by $\bar{\omega}_2$. In contrast, $S(\vec{Q},\omega)$ has a much sharper zero sound spectrum peaked at a frequency $\bar{\omega}_1$. From neutron scattering studies done over the last 20 years at Chalk River, we can say that $\bar{\omega}_1$ exhibits the characteristic phonon-roton dispersion relation, although it is fairly strongly damped. This $\bar{\omega}_1$ mode is very similar to the density fluctuation spectrum of normal liquid ^{3}He. When we go below T_λ, the finite value of $< \hat{\psi}(\vec{r}) >$ leads to the spectra of $G(\vec{Q},\omega)$ and $S(\vec{Q},\omega)$ being mixed and renormalized. This effect is analogous to two pendulums (of frequency $\bar{\omega}_1$ and $\bar{\omega}_2$) which are coupled by a spring (this is the role of $< \hat{\psi}(\vec{r}) >$). The response function of each pendulum has now two renormalized resonances (denoted by ω_1 and ω_2), although with quite different weights.

In the context of the above picture, a crucial fact about ^{4}He is that the condensate fraction is always less than 9%. The significance of this has been only recently appreciated, namely that we do not expect that the renormalized frequencies (ω_1 and ω_2) to be very different from those <u>above</u> T_λ. Thus we have a natural explanation of the otherwise puzzling experimental fact that the zero-sound resonance frequency ω_1 does not change that much as we go from, say, $1K$ to $2.5K$. Moreover, we also see why the ω_1 zero sound density fluctuation plays the role of the dominant elementary excitation <u>below</u> T_λ, but not above. The reason is that below T_λ, the ω_1 mode appears in the $G(\vec{Q},\omega)$ spectrum with a weight related to $< \hat{\psi}(\vec{r}) >= \sqrt{n_o}$. The original broad spectrum of $G(\vec{Q},\omega)$ associated with $\bar{\omega}_2$ is "swamped" by the increasingly sharp peak at ω_1.

As I have suggested in my Banff article [1], the damping of the ω_1 mode sharply decreases as $T \to 0$ because it is increasingly coupled to zero-sound ω_1 modes rather than the broad ω_2 spectrum. For illustrative purposes, let us consider decay processes. Above T_λ, the decay channel of the zero sound mode might be described by

$$\bar{\omega}_1(\vec{Q}) = \bar{\omega}_2(\vec{p}) + \bar{\omega}_2(\vec{Q} - \vec{p}) \ , \tag{1}$$

while at temperatures somewhat below T_λ, this is replaced by

$$\omega_1(\vec{Q}) = \omega_1(\vec{p}) + \omega_1(\vec{Q} - \vec{p}) \ . \tag{2}$$

The phase space for the decay process (2) is much more restricted than for (1) because of the sharpness of the ω_1 modes. This seems to be a natural explanation of why the zero sound phonon-roton excitation sharpens up so dramatically as we pass into the superfluid phase of liquid ^{4}He.

A good theory must also make new predictions which can be verified experimentally. In this regard, the most interesting possibility would be to find some evidence of the ω_2 mode in the density fluctuation spectrum. According to the preceeding analysis, $S(\vec{Q},\omega)$ should exhibit some structure associated with what one might call the non-zero sound spectrum, with weight roughly proportional to the magnitude of n_0. In order to stimulate an experimental search, GRIFFIN and PAYNE [7] have recently presented some model calculations of $S(\vec{Q},\omega)$ based on the somewhat ad hoc assumption that $\bar{\omega}_2 \simeq Q^2/2m^*$, where m^* was taken to be three times the bare mass m of a ^{4}He atom. SVENSSON and TENNANT [8] have looked for such a mode at $Q^2/2m$ for $Q \simeq 1\text{\AA}^{-1}$ and, with a sensitivity of about 0.1% of the maxon resonance, found nothing. It would be very worthwhile to redo such experiments at somewhat greater sensitivity and to examine a wider region of frequency space below the phonon-roton dispersion curve. We once again emphasize that detection of such low frequency structure in $S(\vec{Q},\omega)$ would be a direct confirmation of the role of the Bose broken symmetry in the dynamics of superfluid ^{4}He.

In their original paper, GRIFFIN and PAYNE [7] suggested experiments looking for the ω_2 mode in $S(\vec{Q},\omega)$ should be done at as low a temperature as possible, to eliminate scattering from thermally excited phonon-roton modes. However, recent work reported by WILLIAMS at this Symposium suggests that the region closer to T_λ might be more relevant, for reasons I will now briefly discuss. In the picture suggested by Williams, while the dominant elementary excitations at low temperatures are still the phonon-roton modes, as the temperature approaches T_λ, the energy of vortex-rings decreases so that they take over as the important fluctuations in the critical region. Williams introduces a Kosterlitz-Thouless type argument to understand how it becomes energetically favourable to excite vortex rings of increasingly large radius and how the bare superfluid density is thus renormalized to zero by the excitation of such soft modes near T_λ.

What we would like to suggest is that at a microscopic level, these vortex-rings are perhaps most naturally understood as the ω_2 poles of the field fluctuation spectrum [i.e., resonances of $G(\vec{Q},\omega)$]. We recall that in a Bose-condensed system, the single-particle Green's functions may be related to the superfluid velocity correlation function $\chi_{v_s v_s}$ [5,9]. Moreover, in the hydrodynamic low frequency region, the ω_2 mode corresponds to second sound, which is usually identified as the Goldstone mode associated with the macroscopic wave function describing the Bose broken symmetry. The suggestion that low energy vortex-ring excitations are the dominant poles of $G(\vec{Q},\omega)$ near T_λ seems quite attractive but at the present time, this must be viewed as speculative. According to our preceding arguments, however, one experimental consequence would be that, however, these vortex-ring excitations should show up as soft modes in $S(\vec{Q},\omega)$, that is, with a weight <u>and</u> frequency which goes to zero as $T \to T_\lambda$. In this connection, it is interesting to recall that according to the numerical results of CEPERLEY and POLLOCK [3], the condensate fraction n_0 drops to zero at T_λ very abruptly. Thus the weight of such vortex-ring excitations in $S(\vec{Q},\omega)$ might be significant in a reasonably large temperature interval close to T_λ.

A separate question concerns what specific evidence can one obtain about the role of the Bose broken symmetry from careful measurements of the zero-sound mode in $S(\vec{Q},\omega)$ as $T \to T_\lambda$. The major difficulty here is that one must first be able to separate the phonon-roton resonance from the "multiphonon" contribution. This problem is discussed in more detail by SVENSSON in this Symposium. At low Q [$\lesssim 0.5 \text{Å}^{-1}$], this is easily done since the multiphonon structure is well separated from the 1-phonon resonance [10]. It would be very useful to carry out a systemmatic experimental study of the temperature dependence of this multiphonon peak as $T \to T_\lambda$ and in the normal phase. For Q in the maxon region [$Q \sim 1 \text{Å}^{-1}$], which has been extensively studied, the multiphonon spectrum strongly overlaps on the 1-phonon resonance and it seems difficult to separate the two contributions with any precision. Coming to even larger Q values [$Q \sim 2 \text{Å}^{-1}$, the roton region], the multiphonon region is reasonably distinct, occurring at higher frequencies than the roton resonance.

At the present time, I feel that the best procedure is to estimate the multiphonon part of the spectrum at low temperatures (where the zero sound 1-phonon resonance is very sharp) and subtract this from $S(\vec{Q},\omega)$ at higher temperatures on the assumption that it doesn't change much with temperature [1,11]. As I have discussed in my Banff review, there is no theoretical justification for the two-component fitting procedure introduced by WOODS and SVENSSON [12]. The results based on such a data analysis should be viewed accordingly. We remark that in the Woods-Svensson interpretation of the temperature dependence of $S(\vec{Q},\omega)$ in the superfluid phase, the weight of the phonon-roton mode in $S(\vec{Q},\omega)$ vanishes above T_λ, even though the frequency of this mode shows very little change as we approach T_λ. This is in sharp contrast with the behaviour of soft modes associated with a broken symmetry. If taken seriously, the Woods-Svensson picture would seem to imply that the phonon-roton resonance still exists above T_λ but is no longer a density fluctuation.

Coming back to predictions about the phonon-roton resonance, we note that there are still very few careful studies of the phonon linewidths in the intermediate temperature region (say $T \gtrsim 1.5°K$). What one is looking for is evidence that the condensate fraction is decreasing. TALBOT and GRIFFIN [11] have predicted that at for small Q, the phonon linewidth should go as

$$\Gamma_Q(T) = A(Q)\frac{\rho_N}{\rho} - B(Q)\frac{\rho_N}{\rho}\frac{\rho_s}{\rho} \ . \tag{3}$$

Here the first term is due to scattering from thermally excited rotons and is the kind of term that one expects in the Landau-Khalatnikov kind of calculation. The second term, however, is directly associated with the existence of a condensate. Several approximations were made in arriving at this particular form (for details, see [11]). We mention it as an example of a condensate-related contribution which does not arise in a picture based on a gas of weakly interacting quasiparticles and hence is of special interest.

3. EXCITATIONS IN A BOSE GAS

It is interesting to contrast the case of superfluid ^4He with a weakly interacting dilute Bose gas (WIDBG). There are several possible systems where a WIDBG may be produced and experimental work in this area is being actively pursued at the present time. Three examples are:

    (a) Spin polarized atomic Hydrogen in magnetic traps [13].
    (b) Atoms trapped by lasers.
    (c) Excitons in optically pumped Cu_2O [14].

In the first two examples, the Bose-Einstein condensation transition may be less than $1mK$ because of the low density; while in the exciton case, T_λ may be as high as $100K$ because of the electronic masses involved. Somewhat surprisingly, the dynamics of a WIDBG at finite temperatures (i.e., with significant depletion of the condensate) was only recently worked out by PAYNE and GRIFFIN [6]. I shall now briefly summarize their results.

In contrast with the Bose liquid discussed in Section 2, $G(\vec{Q},\omega)$ has a <u>sharp</u> free-particle-like pole given by $\bar{\omega}_2 \simeq Q^2/2m$ for $T > T_\lambda$. On the other hand, $S(\vec{Q},\omega)$ has now only a broad spectrum. No zero-sound mode exists because the density is too small to sustain a well defined self-consistent-field. When we go below T_λ, the order parameter $< \hat{\psi}(\vec{r}) >= \sqrt{n_0}$ couples $G(\vec{Q},\omega)$ and $S(\vec{Q},\omega)$ with the result that the renormalized pole of $G(\vec{Q},\omega)$ appears in $S(\vec{Q},\omega)$ with a weight related to n_0.

Because the condensate fraction n_0 of a WIDBG may be close to 100% (at $T = 0K$), the condensate renormalization of $\bar{\omega}_2$ is significant. Surprisingly, the final result of detailed microscopic calculations [6] is that at intermediate temperatures, one has to a good approximation

$$\omega_2(Q) \simeq \left[\left(\frac{Q^2}{2m}\right)^2 + c_0^2 Q^2\right]^{\frac{1}{2}}$$
$$\simeq c_0 Q \text{ for } Q \to 0 \ , \tag{4}$$

where the phonon velocity is given by

$$c_0^2(T) = \frac{n_0(T)V_0}{m} \ . \tag{5}$$

Here $n_0(T)$ may be taken as the temperature-dependent condensate fraction of an ideal Bose gas. We remark that this result reduces to the Bogoliubov result at $T = 0$ but that the standard practise of using this expression with n_o replaced by n is quite incorrect at any finite temperature. Apart from this depletion of the condensate, the main effect of the increasing number of excited atoms outside the condensate is to give the ω_2 mode a <u>finite</u> width which is proportional to the temperature [6].

As we have noted above, this Bogoliubov phonon mode will appear as a resonance in $S(\vec{Q},\omega)$ with a weight $\sim n_0(T)$. Thus we can study the elementary excitations of a WIDBG through $S(\vec{Q},\omega)$ probes once again - but for a completely different reason than in superfluid ^{4}He. In a Bose-condensed gas, the Bogoliubov phonon mode ω_2 is not a zero-sound density fluctuation but the Goldstone symmetry-breaking mode. Its weight in $S(\vec{Q},\omega)$ disappears above T_λ, where it reduces to a free-particle-like excitation.

The dynamics of a Bose-condensed gas are also quite different from a liquid in the low frequency, hydrodynamic region [15]. First and second sound both involve density fluctuations and hence will be both strongly exhibited in $S(\vec{Q},\omega)$ for a WIDBG. This contrasts with superfluid ^{4}He, where second sound is essentially uncoupled from density fluctuations.

ACKNOWLEDGEMENTS

I would like to thank Prof. G. Williams for useful discussions concerning his vortex-ring model and Prof. K. Ohbayashi for making it possible for me to attend the Hiroshima Symposium on Excitations in Quantum Liquids. This work was supported by the Natural Sciences and Engineering Research Council (NSERC) of Canada.

REFERENCES

1. A.Griffin, Proceedings of the Banff International Conference on Quantum Fluids and Solids (October,1986), Can. J. Phys. 65, 1368(1987).

2. V.F. Sears, E.C. Svensson, P. Martel and A.D.B. Woods, Phys. Rev. Letters 49,279 (1982).

3. D.M. Ceperley and E.L. Pollock, Phys. Rev. Letters 56, 351 (1986).

4. E.L. Pollock and D.M. Ceperley, Phys. Rev. B 36, 8343 (1987).

5. P.C. Hohenberg and P.C. Martin, Ann. Phys. (N.Y.) 34, 291 (1965).

6. S.H. Payne and A. Griffin, Phys. Rev. B32, 7199 (1985).

7. A. Griffin and S.H. Payne, J. Low Temp. Phys. 64, 155 (1986).

8. E.C. Svensson and D.C. Tennant, Proc. 18th International Conference on Low Temperature Physics, Kyoto, 1987, in Japan Journ. Applied Physics, 26 Supp 26-3, 31 (1987).

9. J.W. Kane and L.P. Kadanoff, Phys. Rev., 155 80 (1967).

10. E.C. Svensson, P. Martel, V.F. Sears and A.D.B. Woods, Can. Journ. Phys. 54, 2178 (1976).

11. E. Talbot and A. Griffin, Phys. Rev. B29, 2531 (1984).

12. A.D.B. Woods and E.C. Svensson, Phys. Rev. Letters 41, 974 (1978).

13. H.F. Hess, G.P. Kochanski, J.M. Doyle, N. Masukara, D. Kleppner and T.J. Greytak, Phys. Rev. Letters, 59, 672 (1987).

14. D. Snoke, J.P. Wolfe and A. Mysyrowicz, Phys. Rev. Letters 59, 827 (1987).

15. C. Gay and A. Griffin, J. Low Temp. Phys. 58, 479 (1985).

Part II

Experimental Study of Elementary Excitations: Raman Scattering and Neutron Scattering

Raman Scattering of Liquid ^{4}He

K. Ohbayashi

Faculty of Integrated Arts and Sciences, Hiroshima University,
Hiroshima 730, Japan

1. Introduction

In our recent measurements of Raman scattering from superfluid ^{4}He [1–6], we have found several new results which in one part verify theoretical predictions left untested and in another part demand re-investigation of seemingly well established concept of two–roton bound state. HALLEY discusses theoretical problems in this volume, and we describe experimental results in this report.

How the scattered light spectrum of liquid ^{4}He appears is briefly described with an extremely simplified illustration of the experimental set up shown in Fig 1. Liquid helium is illuminated by a laser beam. Scattered light is analyzed by a spectrometer. On the exit focussing plane of the spectrometer, the spectrum of the scattered light appears. $\vec{q} = \vec{k}_i - \vec{k}_0$ is called scattering vector and is characteristic quantity of scattering. Here, $\vec{k}_i$ and $\vec{k}_0$ are wavevectors of incident and outgoing light,respectively. With scattering angle θ; $q = 2k_i\sin(\theta/2)$. The spectrum is plotted as a function of energy shift from the incident laser light and has three components ; Rayleigh scattering, Brillouin scattering, and Raman scattering.

Rayleigh scattering and Brillouin scattering are often called quasielastic light scattering or the central component. Quasielastic light scattering occurs in first order and its scattering mechanism under usual conditions is well known [7–11]. The scattering rate of the central component is given by

$$h_1(\omega) \propto \int dt \int d\vec{r}_1 \int d\vec{r}_2 \alpha_0^2 (\hat{\epsilon}_0 \cdot \hat{\epsilon}_n)^2 \langle \rho(\vec{r}_1,0)\rho(\vec{r}_2,t) \rangle e^{-i\omega t + i\vec{q}\cdot(\vec{r}_1 - \vec{r}_2)}, \tag{1}$$

where α_0 is the atomic polarizability and $\rho(\vec{r},t)$ is the density at position $\vec{r}$ and at time t. $\hat{\epsilon}_0$ and $\hat{\epsilon}_n$ are respectively the polarizations of the incoming and outgoing light in the scattering process. α_0 is well known and the two–body correlation can be related to thermodynamic quantities using hydrodynamics and thermodynamics. The intensity and spectral shape can be calculated easily. Rayleigh scattering detects density fluctuations coupled with entropy fluctuations. In normal fluid ^{4}He,it is diffusive and Rayleigh spectrum is a lorentzian with no energy shift from the incident light and with width $\Gamma_R = D_T q^2$ proportional to the thermal diffusivity D_T. In superfluid ^{4}He, entropy fluctuations propagate as second sound with velocity c_2 and give doublet lorentzian peaks at energy shift $\pm c_2 q$. The width is proportional to the damping of the second sound. Because the coupling of the entropy fluctuation to the density fluctution is very weak in superfluid ^{4}He at saturated vapor pressure, the doublet is hardly visible. Brillouin scattering detects the density fluctuation coupled with first sound wave of velocity c_1. It shows a doublet–lorentzian spectrum at energy shift of $\pm c_1 q$

Springer Series in Solid-State Sciences, Vol. 79 **Elementary Excitations in Quantum Fluids**
Editors: K. Ohbayashi · M. Watabe © Springer-Verlag Berlin, Heidelberg 1989

with width Γ_B proportional to the damping coefficient of first sound wave. As thus described, the central component is well understood and we use it as a reference of Raman scattering. However, it is noted that, near a second order phase transition such as the gas-liquid critical point [12] or lambda point[13], the spectum of the central component does not take the form described here. And it gives us unique information on critical fluctuations that cannot be obtained by other methods.

The Raman spectrum is observed in an energy shift region larger by a factor of about 10^2 and with intensity weaker by $10^{-4} - 10^{-5}$ compared with the central component. In normal fluid ^{4}He, it shows a weak peak at energy shift of about 25 K and then decreases nearly exponentially as the energy shift increases as shown in (a) of Fig. 1. If the temperature is cooled down and liquid ^{4}He becomes superfluid, the spectrum shows a drastic change as shown in (b) of Fig. 1. Two peaks, p_1 and p_2, and other fine structures appear. In 1969, HALLEY[14] proposed that Raman scattering could be used to study the elementary excitations in superfluid ^{4}He. Neutron inelastic scattering is a direct method to determine the dispersion relation of the elementary excitation. It measures the dynamical structure factor $S(k,\omega)$ of superfluid ^{4}He as a function of wave-vector k and energy ω. A typical ω - dependence of $S(k,\omega)$ at fixed k is schematically shown in Fig. 2 (a) with a broken line. By plotting the strongest peak position on the k-ω plane, the phonon-roton dispersion curve of the elementary excitaion is obtained as shown by the solid curve. Excitations near the maximum region and those near the terminal region are often called maxons and plateaus, respectively. We denote the energy of the local minimum by Δ_0, the local maximum by Δ_1, and the terminal plateau by Δ_2. Because the scattering vector of light $q \simeq 10^{-4}\overset{o}{A}^{-1}$ is much smaller than the wave-vector of a roton, maxon, or plateau, light cannot excite these elementary excitations in a first order process. Halley's proposal was a second order scattering process of simultaneous creation of two elementary excitations of wavevectors almost equal in magnitude and directed oppositely. He expected three peaks at $2\Delta_0$, $2\Delta_1$, and $2\Delta_2$, where divergence of the density of state is expected. The first suc-

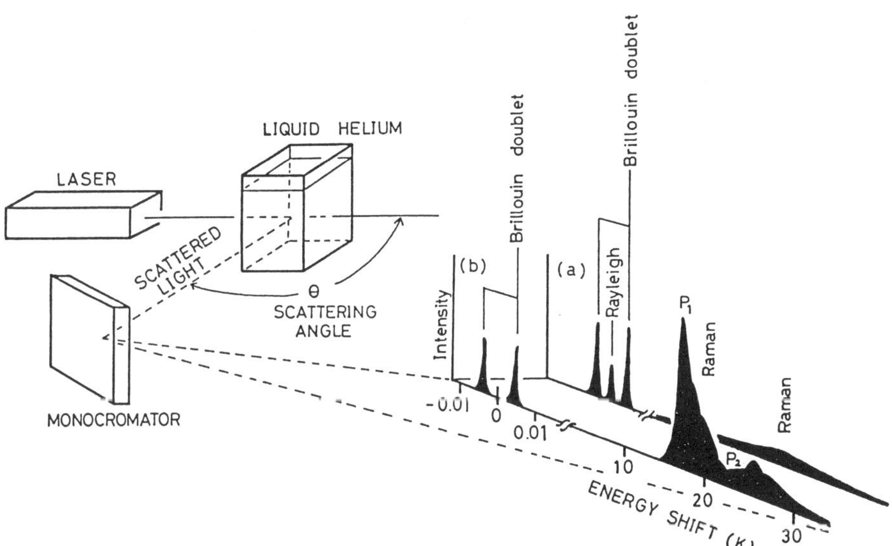

Fig. 1.   Experimental method of light scattering from liquid helium.

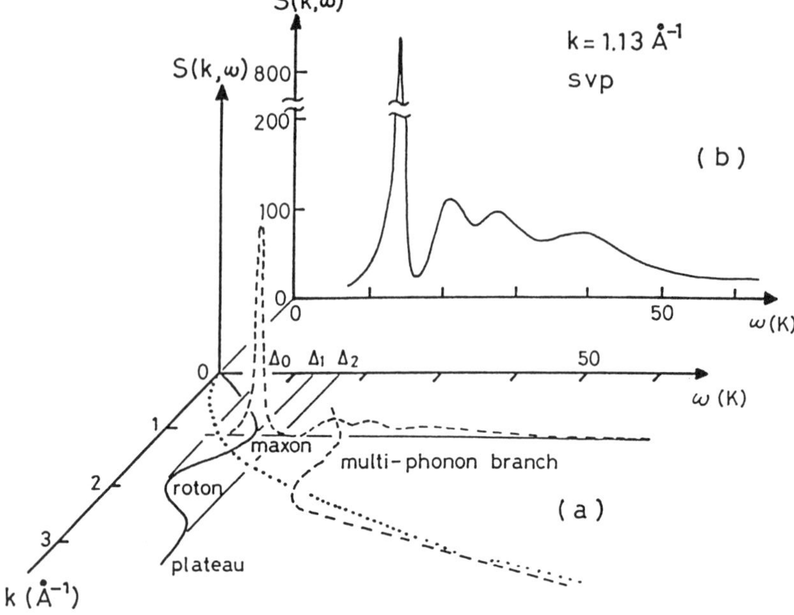

Fig. 2.(a) Dynamical structure factor $S(k,\omega)$ of superfluid ⁴He plotted as a function of wavevector k and energy ω. The vs k relation of positions of the strongest peak gives the phonon–roton dispersion curve. (b) In the upper branch, three peaks have been confirmed by STIRLING [29]. A schematic hand drawing of his data is shown here.

cessful experiment by GREYTAK and YAN [15] verified this prediction by observing p_1 near $2\Delta_0$. Subsequently, the theoretical prediction was made by IWAMOTO [16] and RUVALDS and ZAWADOWSKI [17] that the two rotons excited by the light make a bound state. The GREYTAK group [18,19] experimentally observed the spectral shape consistent with the theoretical prediction. The agreement between the experiment and the theory being rather good, the concept of the bound state was widely accepted as an established fact. Experimentalists left this field with the result that no experimental report of Raman scattering has been made since 1976 until quite recently our group began new experiments.

Meanwhile, however, theoretical works were continued by the Minnesota group [20–26] as explained by HALLEY and CAMPBELL in this volume. Following Halley's expression the Raman scattering rate $h_2(\omega)$ may be written in the form

$$h_2(\omega) \propto \int dt \int d\vec{r}_1 \int d\vec{r}_2 \int d\vec{r}_3 \int d\vec{r}_4 \, \hat{\epsilon}_0 \cdot \alpha(\vec{r}_1,\vec{r}_2) \cdot \hat{\epsilon}_n \hat{\epsilon}_0 \cdot \alpha(\vec{r}_3,\vec{r}_4) \cdot \hat{\epsilon}_n$$

$$\times \langle \rho(\vec{r}_1,0)\rho(\vec{r}_2,0)\rho(\vec{r}_3,t)\rho(\vec{r}_4,t)\rangle e^{-i[\omega t - \vec{k}_i \cdot (\vec{r}_2-\vec{r}_4) + \vec{k}_0 \cdot (\vec{r}_1-\vec{r}_3)]}. \quad (2)$$

Here, $\alpha(\vec{r},\vec{r}')$ is a polarizability tensor. The scattering formula of Eq.(2) is well established. However, to calculate the Raman spectrum is not easy. One reason is that the polarizability tensor $\alpha(\vec{r},\vec{r}')$ is not well known. The Minnesota group investgated it theoretically and predicted existence of s–

wave scattering quantitatively. Another reason is that an exact relation between the density fluctuation and the elementary excitation is not known. RUVALDS and ZAWADOWSKI assumed a linear relation and derived the concept of two-roton bound state. However, KLEBAN and HASTINGS [23] proved that the linear relation contradicts excluded volume conditions. This suggested the need for experimental re-investigation of the two-roton Raman scattering. The relation between density fluctuations and the boson operator of the elementary excitation and the relation between the four-body correlation function and the two-body correlation function are fundamental problems. Detailed study of Raman scattering will give useful clues to clarify them. Although our first motivation to re-start the experiments was to examine fine structures which the present author detected more than ten years ago [27], we also intended experimental test of the theoretical predictions by the Minnesota group. Because Raman scattering is complementary to neutron scattering, we also focussed our attention on current problems of the dynamical structure factor $S(k,\omega)$ that can be investigated also by Raman scattering. In this report, we discuss the following problems:

(a) Pressure dependence of two-roton scattering:
We measured the pressure dependence of the two-roton Raman scattering and found that the shift of the peak from $2\Delta_0$, twice the energy of the single roton determined by the neutron scattering, changes from negative at low pressure to positive at higher pressure. The spectral shape at higher pressure cannot be explained by the two-roton bound state model.

(b) Polarization analysis:
To separate s-wave spectrum from d-wave spectrum, polarization analysis of the scattered light was made. s-wave was detected as predicted by the Minnesota group. This shows that a simple calculation by a dipole-induced-dipole model is not adequate.

(c) Fine structure:
In addition to the two-roton Raman scattering, a few fine structures were confirmed. We discuss the structure as Raman processes associated with more than two elementary excitations, theoretically suggested by IWAMOTO's group recently [28], and in relation to the structures in the upper branch of $S(k,\omega)$ recently found in neutron scattering [29].

(d) Temperature dependence:
Determination of the roton energy by neutron scattering is model dependent and does not give a definite result[30]. We discuss the problem in relation to temperature dependence of the two-roton peak. The two-component model of $S(k,\omega)$ proposed by WOODS and SVENSSON [31] is also discussed.

2. Experimental Apparatus

The block diagram of the experimental system is shown in Fig.3. Incident light source was an argon ion laser operated at 5145 $\overset{\circ}{A}{}^{-1}$(Nippon Electric Company model no. GLG3200). The laser beam out of the source was vertically polarized and a half-wave plate was used as the polarization rotator. The polariser 1 was used to determine the direction of the polarization of the incident light accurately. The laser beam was focussed into sample with the lens 1.The sample was cooled either with a pumped ^{4}He type cryostat down to 1.3 K or a ^{3}He circulation type cryostat down to 0.6 K. The temperature of

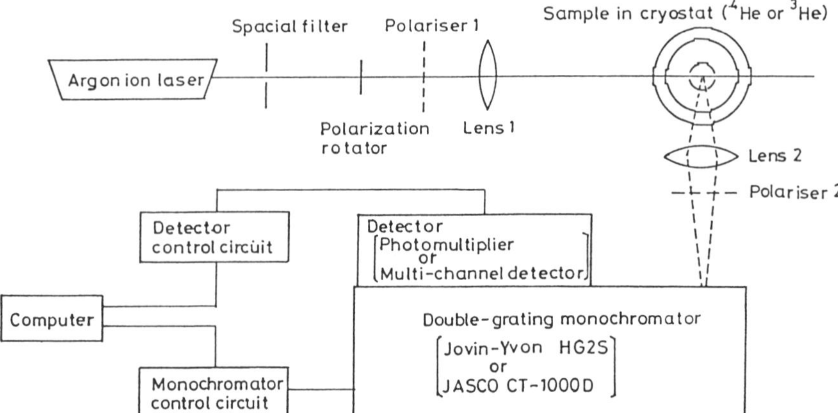

Fig. 3. Block diagram of the experimental system for light scattering from liquid helium. For details see text.

the sample was measured by a germanium resistor and a precision pressure gauge (Texas Instrument Inc. model no. 145). The typical value of the laser power used was 800 mW with the ^{4}He cryostat and 400 mW with the ^{3}He cryostat. Scattered light was collected at 90^0 to the incident beam with the f/3.7 lens 2. The analyzing polariser 2 was used with its easy axis vertical. To measure a spectrum, either a single channel method or a multichannel method was used. In both methods, a computer was used to control the spectrometer scanning and the data acquisition procedure.

In the case of the single channel method, a Jovin-Yvon model no. HG2S double grating monochromator and a Hamamatsu Photonics model no. R464 photomultiplier tube were used. For amplification and discrimination of output pulses from the photomultiplier tube, a Princeton Applied Research model no. 1121 photon counter was used. The dark count rate of the photomultiplier tube was less than 0.7 counts/sec.

For the multichannel detection, a Jasco CT1000D type double grating monochromator with a Princeton Instruments model IRY-700 multichannel detector were used. Compared with the single channel method, a considerable improvement in signal to noise ratio has been achieved by adopting the multichannel detector, with which the intensity at 700 different energy shift values were measured simultaneously. The energy shift per channel was 0.15 cm^{-1}. The detector was exposed to the scattered light spectrum during a 700 sec. interval after which the data were transferred to a microcomputer. To get good statistical accuracy, we needed an accumulation of about 50 such exposures. The background noise observed for the detector in the dark was also accumlated more than 80 times with the same exposure time. It was suitably normalized and was subtracted from the observed signals. The sensitivities are slightly different at different channels. For their calibration, the spectrum of an incandescent bulb was measured. We regarded it essentially white in the spectrum range concerned. Because the exit slit was wide open the tail of the central commponent continued to the two-roton peak region with typical contribution of about 10 % of the two-roton peak height. The functional form of this stray light was exponential and decreased rapidly with an increase of energy shift. It was carefully determined and subtracted from the data. The calibration of the wavevector was done using a mercury discharge lamp and a few emission lines of the laser. We estimate the absolute accuracy of determination of the energy shift from the central peak to be $\pm$ 0.2 K.

In this paper, we report mainly four series of measurements. The two-roton spectra were measured with the ^{3}He cryostat and the multichannel detector, the polarization properties with the ^{4}He cryostat and the single channel detector, the fine structures with the ^{4}He cryostat and the multichannel detector, and temperature dependence with the ^{4}He cryostat and the multichannel detector. Except for the polarization measurement, the polarization rotator, the polariser 1, and the polariser 2 were not used.

3. Two-Roton Raman Scattering

In Fig. 1, the sharp peak p_1 is the two-roton Raman scattering peak. We measured pressure dependence of the two-roton peak at a temperature of 0.65 K. Fig. 4 shows how the two-roton peak changes with pressure. With an increase of pressure, both the intensity and the energy shift of the peak decrease. However, the shift does not exactly follow the pressure dependence of the roton energy Δ_0 measured by neutron scattering. The two-roton peak appears at an energy lower than $2\Delta_0$ at svp but shifts to an energy higher than $2\Delta_0$ at pressures above about 2 kg/cm^2. The positive deviation increases with an increase of pressure.

The two-roton Raman intensity is mainly proportional to the density of state $D_2(\omega)$ of the two rotons created by light. RUVALDS and ZAWADOWSKI [17] theoretically described (the RZ theory) the density $D_2(\omega)$ of two attractively interacting rotons with total momentum zero as

$$D_2(\omega) \propto 2\pi E_i^{\frac{1}{2}}\delta(\omega-2\Delta_0+E_i)+(\omega-2\Delta_0)^{\frac{1}{2}}(\omega-2\Delta_0+E_i)^{-1}\theta(\omega-2\Delta_9). \tag{3}$$

Here, $\theta(x)$ is a unit function which is one for $x>0$ and zero for $x<0$. The δ-function represents a single bound state with binding energy E_i for attractive roton-roton interaction. The second term represents unbound states. The first term gives a sharp peak with negative shift and the second term gives a broad peak with positive shift. If the interaction between rotons is repulsive, the bound state disappears and the density of unbound states is given by the same expression as the second term of Eq. (3). In this case E corresponds to the repulsive energy between two rotons. For $E_i << \omega -2\Delta_0$, the density approaches to that for non-interacting rotons; $D_2(\omega) \propto (\omega-2\Delta_0)^{-\frac{1}{2}}$ $\times\theta(\omega-2\Delta_0)$. In addition to the main factor $D_2(\omega)$, the k- dependence of a few other factors also affects the Raman spectral shape. These are the coefficient of four-body interaction between quasiparticles $g_4(k)$, the light-helium coupling constant $t(k)$, and the proportionality factor $Z(k)$ between density fluctuations and boson operators. However, if a comparison is restricted to a narrow energy region, these effects are very weak and may be neglected.

The very sharp, negatively shifted two-roton Raman spectrum was first measured by the GREYTAK group at svp [18,19]. The agreement between the experimental spectrum and the theoretical curve of Eq. (3) was good enough for them to conclude that the negative shift is due to an attractive interaction between two rotons created by light and the sharp peak is evidence of the two-roton bound state. They also claimed[19] that the fit was excel-

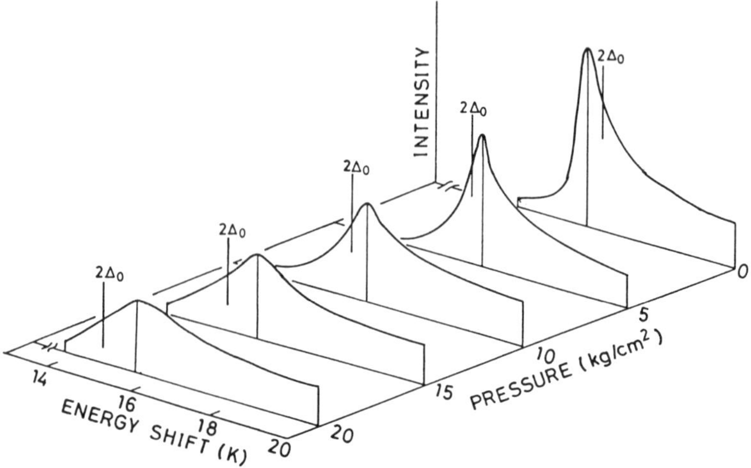

Fig. 4.   Pressure dependence of the two-roton Raman scattering measured at a temperature of 0.65 K with an instrumental FWHM of 0.9 K. The curves are hand drawing through the data points. $2\Delta_0$ indicates twice the energy of a single roton determined by neutron scattering.

lent enough to determine both the roton energy and the binding energy simultaneously without recourse to the data of neutron scattering.

As shown in Fig. 4, we observed that the shift of the two-roton peak from $2\Delta_0$ changes from negative to positive as pressure increases. It is interesting to test whether all these spectra are consistently fitted by the RZ theory. Our svp spectrum can be fitted by the theory reasonably well as the Greytak group found. At 20 kg/cm² ,however, the peak shows a relatively large positive shift of 1.0 ± 0.2 K. Within the framework of the RZ theory, this spectrum must be analyzed by the repulsive interaction model,i.e. the second term of Eq.(3).Although this term shows a maximum at $2\Delta_0 + E_i$,we cannot relate the observed shift to E_i directly. In an actual experiment, finite instrumental resolution function broadens the spectrum and the peak position shifts accordingly. Then, the result of a convolution integral of the theoretical function with the instrumental resolution profile function must be compared with experimental result. The line shape of the central component, which in our case can be regarded as the instrumental profile,is plotted by an alternate long and short dash line in Fig.5. The instrumental function may be approximated as a lorentzian with a half width at half maximum of 0.35 K. The only adjustable parameter is the repulsive energy E_i and the shift of the peak from $2\Delta_0$ changes with the choice of E_i. And the observed shift of 1.0 ± 0.2 K is expected at E_i = 0.4 ± 0.05 K. The theoretical curve of this parameter is compared with the observed spectrum in Fig. 5. The very poor fit shows that the RZ theory cannot explain the shift and the spectral shape observed at the pressure of 20 kg/cm² consistently. The observed spectrum is much more sharp compared with the theoretical curve. Especially on the lower energy side, the experimental data closely follow along the instrumental profile. The fact suggests that the spectrum shows step like change just at the peak position or has a δ-function peak. As shown in Fig. 4, a positively shifted sharp peak is observed at pressure above about 2 kg/cm². Then, the sharp peak is not neccesarily the evidence

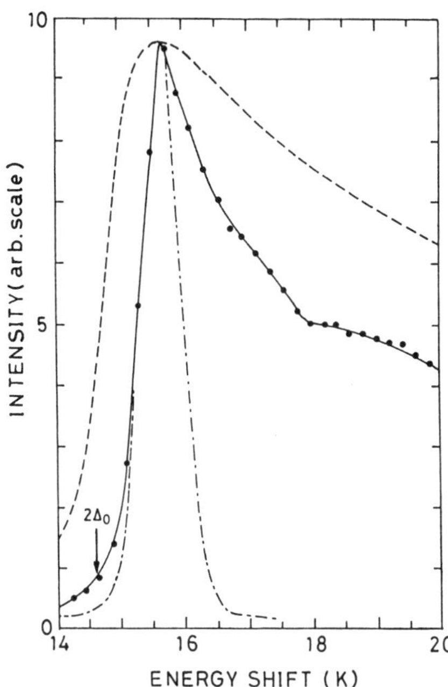

Fig. 5. The two-roton Raman spectrum of superfluid ^{4}He measured at a temperature of 0.65K and at a pressure of 20 kg/cm^2. The broken curve is a theoretical calculation by the RZ theory for a case of repulsive interaction of 0.4K. The alternate long and short dash curve is the instrumental profile determined by the central component. $2\Delta_0$ shows twice the energy of a single roton.

of the two-roton bound state under an attractive interaction, which should be negatively shifted.

Until quite recently, experimental data that support the RZ theory have been accumulated as explained by RUVALDS in this volume. The theory starts from the quasiparticle Hamiltonian

$$H = \sum_{\vec{q}} \epsilon_{\vec{q}} a^{\dagger}_{\vec{q}} a_{\vec{q}} + \sum_{\vec{q},\vec{k}} g_3(\vec{q},\vec{k})(a^{\dagger}_{\vec{q}} a^{\dagger}_{\vec{k}} a_{\vec{q}+\vec{k}} + h.c.)$$
$$+ \sum_{\vec{q},\vec{k},\vec{l}} g_4(\vec{q},\vec{k},\vec{l})(a^{\dagger}_{\vec{q}} a^{\dagger}_{\vec{k}} a_{\vec{l}} a_{\vec{q}+\vec{k}+\vec{l}} + \cdots). \tag{4}$$

Here, $\epsilon_{\vec{q}}$ is the non-interacting quasiparticle excitation energy and $a_{\vec{q}}$ is the boson operator of quasiparticle. The four-point interaction g_4 describes quasiparticle pair scattering and results in bound state. The three-point interaction term g_3 describes the scattering of one quasiparticle into a quasiparticle pair, and leads to the splitting of the excitation spectrum into two branches due to the hybridization between two-roton bound states and the Feynman-Cohen dispersion curve. To relate the density fluctuation in Eq. (2) to the quasiparticle operator, they used a linear relation

$$\rho_{\vec{k}} - Z(k)^{\frac{1}{2}}(a_{-\vec{k}} + a^{\dagger}_{\vec{k}}) \tag{5}$$

Taking g_4 constant, Eq.(3) is derived. Concerning a few arguments that support the theoretical results of the quasiparticle interaction theory, we refer the reader to the book edited by RUVALDS and REGGE[32]. One result of the theoretical model is the two-roton bound state, which was supported by the Raman scattering results of the GREYTAK group as mentioned above. Another result is a split of $S(k,\omega)$ into two branches due to the hybridization of the Feynman–Cohen dispersion curve with the two–roton bound state. This explained an early neutron observation of $S(k,\omega)$ by COWLEY and WOODS [33]. A broad weak peak was observed on the higher energy side of the single phonon peak. The upper branch was obtained by plotting the mean value of this peak on the k–ω plane as shown by the broken curve in Fig. 2(a).The theoretical model was also used to interpret an apparent contradiction of neutron scattering results to PITAEVSKII's condition [34] that Δ_1 and Δ_2 cannot exceed $2\Delta_0$, twice the energy of single roton. In the experiments at applied high pressure, both the maxon energy Δ_1[35,36] and the plateau energy Δ_2 [37] were observed to exceed $2\Delta_0$. The quasiparticle interaction theory could be used to explain the fact by assuming repulsive roton–roton interaction by SMITH ET AL.[38]. (Note that a temperature effect also plays a role to cause an apparent shift of Δ_2 to higher energy as discussed by NAGAI[39].)

However, there are also arguments against the Ruvalds–Zawadowski theory. As we have just shown, the model does not fit well to our Raman scattering results. The theory was extended by ZAWADOWSKI, RUVALDS, and SOLANA [40] (the ZRS model). HASTINGS and HALLEY [20] theoretically showed that the ZRS model cannot provide a quantitative account of the structure of $S(k,\omega)$. KLEBAN and HASTINGS [23] showed that the two–roton bound state violates excluded volume conditions. Experimentally, SVENSSON ET AL.[37] showed there is no evidence of two–roton bound state in their neutron inelastic scattering data. They also suggested the existence of structure in the multiparticle branch more than ten years ago. And recently, STIRLING [29] clearly observed a few distinct peaks in the multiparticle branch of $S(k,\omega)$. A part of his $S(k,\omega)$ data, at $k = 1.13$ Å^{-1} and at svp, is sketched in Fig. 1 (b) as a function of ω. The curve is an approximate hand drawing after his data and original paper should be referred to for accuracy. The data shows there are three peaks in the upper branch of $S(k,\omega)$,which therefore does not have a simple two–branch structure. MANOUSAKIS and PANDHARIPANDE[41] showed that fine structures appear in the multiparticle branch by developing the perturbation theory in the Feynman–Cohen basis. The theoretical $S(k,\omega)$ is qualitatively similar to the experimental results by STIRLING. In their calculation, they considered not only roton–roton interaction but also roton–maxon and maxon–maxon interaction. Having this theoretical and experimental evidence, it is definitely necessary to re–consider the theoretical framework. Possible directions of future theoretical works are discussed by HALLEY in this volume.

Concerning the difference between the two–roton peak position and $2\Delta_0$ observed by the neutron scattering, we mention the work by BEDEL, PINES,and ZAWADOWSKI [42]. Introducing a pseudopotential for the roton–roton interaction, they theoretically derived properties of superfluid ^{4}He at finite temperatures for various applied pressures. One of their predictions was that the roton–roton interaction observed by Raman scattering (d–wave, $q\approx0$) would change from attractive at low pressures to repulsive at higher pressures. As shown in Fig.4, our data show that the peak position of the two–roton Raman scattering shows a negative shift at low pressures and changes to a positive shift at higher pressures from the energy $2\Delta_0$ equal to twice the single roton energy determined by neutron scattering. The negative (positive) shift means less (more) energy is needed to create a pair of rotons than to make two separated single rotons.It is natural to relate this shift to the interaction between two rotons. Then our data indicate that roton–roton interaction changes from attactive to repulsive as the pressure increases, which is qualitatively consistent with their theory.

Although our results presented here are clear enough to make some conclusion, we like to mention that measurements with better instrumental re-

solution and accuracy are necessary to make a more definitive discussion.
Such work is now being done by us and the results will be reported soon
elsewhere.

4. Polarization and Intensity

In order to calculate the Raman spectrum of superfluid ^{4}He using Eq.(2),
functional forms of the polarizability tensor $\alpha(\vec{r}_1,\vec{r}_2)$ and the four-body
correlation function of density fluctuations must be known. Although both
of them are not well understood, useful information to interpret the polarization tensor can be obtained by experimental measurement of the intensity and polarization property. Symmetry of the polarizability tensor can be
determined by the polarization analysis and spacial dependence of the tensor sensitively affects the intensity of the spectrum.

In Fig. 6, atomic contributions to light scattering process are shown
schematically. In the case of quasielastic light scattering shown in Fig.
6 (a), a single atom is involved in the elementary scattering process. The
incident light polarizes each atom. The scattered light is the sum of the
dipole radiations from each polarized atom and its spectrum is given by Eq.
(1). Thus microscopic process is polarization of an atom and fully characterized by the atomic polarizability α_0, which is well known. On the other
hand, Raman scattering occurs via two atoms. The incident light excites the
first atom, the second atom is virtually excited by the first atom through
the inter-atomic interaction, and then the latter atom deexcites accompanying an emission of light. If the two atoms are far apart,the second atom is
excited by the dipole field of the first atom as shown in Fig. 6 (b). This
Raman process is called the dipole-induced-dipole (DID) Raman scattering.
If the two atoms are situated closely and overlap of the electronic wave
functions of the two atoms cannot be neglected, the direct interaction due
to electronic overlap (EO) dominates the scattering, which is shown in Fig.
6 (c). The change from the DID process to the EO process is continuous and
not distinct.

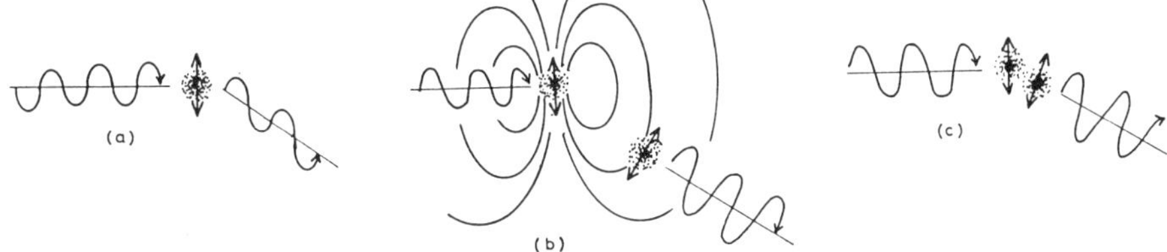

Fig. 6.   Light scattering process. (a) Quasielastic light scattering.
(b) DID Raman process. (c) EO Raman process.

By a kinematical theory [16], the polarization property of the Raman
spectrum $I_R(\omega)$ is given as

$$I_R(\omega) = (\hat{\epsilon}_0 \cdot \hat{\epsilon}_n)^2 I_s(\omega) + [3/4+(\hat{\epsilon}_0 \cdot \hat{\epsilon}_n)^2/4]I_d(\omega). \tag{6}$$

The diagonal part of the polarizability tensor contributes to $I_s(\omega)$, which
is called the s-wave scattering. $I_d(\omega)$ is called the d-wave scattering and
related to non-zero off diagonal part of the polarizability tensor. The DID

process contributes only to the d-wave scattering. The EO process contributes both to the s-wave scattering and the d-wave scattering. Thus the polarization analysis is a useful method to investigate the Raman scattering
process.

STEPHEN [43] developed a theory assuming the DID interaction, which results in a fully d-wave scattering. FETTER [44] postulated weak s-wave contribution to the two-roton Raman scattering. NAKAJIMA [45] pointed out importance of the EO interaction. Detailed experimental polarization analysis
of the two-roton Raman scattering by WOERNER and GREYTAK[46] was consistent
with the d-wave scattering within experimental accuracy. From the fact, the
DID process was believed to be the dominant process. However,later on, KLE-
BAN and HALLEY [24] made a semi-empirical microscopic calculation and found
considerable contribution of the s-wave scattering.They estimated the ratio
of integrated s-wave scattering I_s to the integrated d-wave scattering I_d
to lie within

$$0.11 \leq I_s /I_d \leq 1.45. \tag{7}$$

CAMPBELL, HALLEY, and PINSKI[26] made microscopic calculation and obtained

$$0.08 \leq I_s /I_d \leq 1.88. \tag{8}$$

Because the experimental spectrum near the two-roton region is fully d-wave
scattering, KLEBAN and HALLEY postulated an increase of the s-wave scattering at large energy shift region.

We made careful experimental study of the depolarization ratio of Raman
scattering both from superfluid and normal fluid ⁴He.By observing the spectrum under two   different polarization conditions, we separated the spectrum into s-wave and d-wave using Eq.(6).A result for normal fluid ⁴He at a
temperature of 4.2 K and at a pressure of 2.5 kg/cm² is shown in Fig. 7. As
predicted by KLEBAN and HALLEY,contribution of the s-wave scattering to the

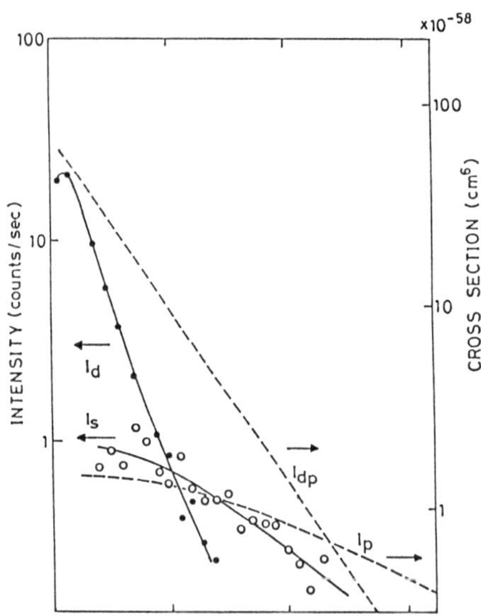

Fig. 7. Polarization property of
the Raman spectrum of normal fluid
⁴He measured at 4.2 K and at 2.5
kg/cm². Closed circles and open
circles, respectively, show d-wave
and s-wave scattering. The solid
curves are drawn through data
points fitted by eye. The broken
lines are theoretical polarized
(I_p) and depolarized (I_{dp}) spectra
of 'gaseous' ⁴He (FROMMHOLD [47]).

42

total scattering increases with an increase of energy shift. By integration
of the spectra over energy shift, we get the experimental ratio $I_s/I_d =$
0.06 - 0.12. We made similar measurements both for superfluid and normal
fluid ^{4}He at various pressures from svp to 20 kg/cm^2.   All of our measured
values of I_s/I_d lie within 0.06 - 0.16. These values are marginally within
the above mentioned theoretical predictions if we consider the experimental
uncertainty. The existence of the s-wave scattering shows that simple DID
approximation is not sufficient to deal with the Raman scattering from liquid helium.

Although the theoretical calculation of the integrated intensity ratio
shows a good fit with experimental results, there is no theory which gives
us explicit functional form of the energy shift dependence of the partial
wave spectra of liquid helium. However, FROMMHOLD and his group [47,48] developed both experimental and theoretical study of the collision induced
Raman scattering from gaseous helium. They observed good fit of their ab
initio calculation to their experimental data. Their theoretical curves
using the HFDHE2 interaction potential [49] of helium are shown with broken
curves in Fig. 7. Their notations I_{dp} and I_p respectively correspond to I_d
and I_s of present work. The vertical position of I_{dp} and I_p is suitably
shifted so that the theoretical I_p curve passes through the experimental
data points of the s-wave scattering. A pair of spectra I_d and I_s has certain similarities to a pair of spectra I_{dp} and I_p.   The energy dependence
of all the spectra is nearly exponential at large energy shift region.   The
slope of I_d (I_{dp}) is steeper than that of $I_s(I_p)$. The intensity ratio I_s/I_d
has a value close to I_p/I_{dp} at small energy shift region. The main difference is noticed in the slope of the exponential energy dependence. In the
case of the collision induced Raman scattering, the slope is related to the
collision time. The longer the collision time is, the steeper the slope is.
The mean speed of the atomic motion is slower in liquid at low temperature
than in gas at higher temperature.   Therefore the mean collision time is
longer in liquid than in gas, which qualitatively explains the steeper
slope of I_d than I_{dp}. Another small difference is appearance of a weak peak
in I_d at small energy shift region, which may be regarded as temperature
broadened reminiscent of the structure observed for superfluid helium. The
overall similarity suggests the possibility of an ab initio calculation of
theoretical s- and d- partial wave scattering spectra. Especially at large
energy shift region, where the inpulse approximation may be applied [50],
effect of density correlations becomes weak. And we may treat the motion of
individual atom in liquid similarly to that in gaseous state.

BAERISWYL [51,52] showed that calculated Raman scattering intensity sensitively changes depending on the choice of the functional form of spacial
dependence of the polarizability tensor.   Thus measurement of the cross
section of Raman scattering is useful to investigate the polarizability
tensor. Because the central Rayleigh-Brillouin scattering intensity I_c can
be calculated accurately, we can use it as a reference and measure the intensity ratio I_R/I_c to estimate absolute Raman cross section. Actual experimental intensity depends on collection optics and efficiency of the detector and the spectrometer. These factors are, however,equivalent both for
the central peak and the Raman spectrum. The main experimental factor that
changes the ratio is the width of the instrumental profile. The ratio increases with an increase of the width, because the Raman spectrum is continuous while the central peak is effectively a line spectrum. From Eq. (1),
the well known formula [7] of Rayleigh-Brillouin scattering is derived as

$$I_c \propto (\omega_i^4 k_B T/6\pi c^4 \rho)[\partial\rho/\partial P]_T (n^2-1)^2 I_i .\tag{9}$$

Here , ω_i is the angular frequency of the incident light, c the speed of light in liquid helium, k_B the Boltzmann constant, T the absolute temperature, ρ the density of liquid helium, P pressure of liquid helium, n the refractive index of liquid helium, I_i intensity of incident light, and m the mass of a helium atom. n is related to the atomic polarizability α_0 by the Lorentz–Lorenz relation

$$(n^2-1)/(n^2+2) = (4\pi\alpha_0\rho/3m). \tag{10}$$

In Fig. 8, pressure dependence of maximum of the central peak intensity and maximum of the Raman intensity are shown. The central peak was measured with an exposure time of 0.32 s and the Raman scattering 700 s. After the correction of the exposure time difference, the intensity ratio is obtained as shown in Fig. 8. The measurement was done at 1.3 K with an instrumental half width at half maximum of 0.8 K. The ratio gradually decreses from $(4.0 \pm 0.4) \times 10^{-5}$ at svp to $(3.3 \pm 0.3) \times 10^{-5}$ at 20 kg/cm^2 as pressure increases. In one of our previous papers, we measured this value with an instrumental full width at half maximum of 3.7 K and observed the ratio of $(2.2 \pm 0.3) \times 10^{-4}$ at all pressures.

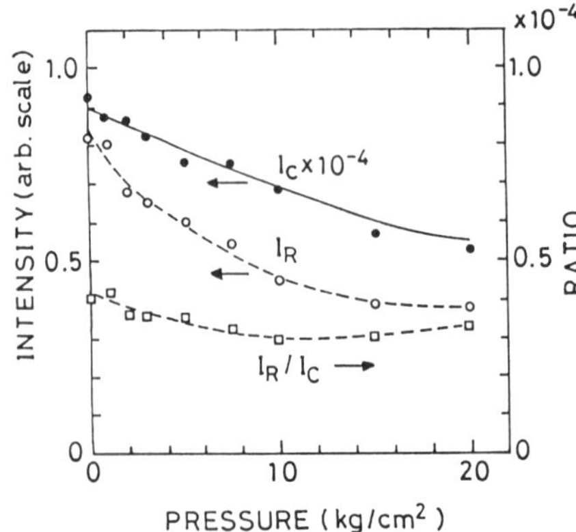

Fig. 8. Pressure dependence of intensities. I_c is the intensity of the central component. I_R is the two–roton Raman scattering intensity. I_R/I_c is the ratio of the two–roton Raman scattering intensity to the intensity of the central component.

We compare observed pressure dependence of the central peak with calculated function using Eq.(9). In the equation, only ρ and $(\partial\rho/\partial P)_T$ are pressure dependent at constant temperature. We can find the values of ρ (g/cm^3) as a function of pressure P (atm) in Table A2 of Ref. [53]. A best fit of these values at 1.3 K to the formula

$$\rho = a_0 + a_1 P + a_2 P^2 + a_3 P^3 \tag{11}$$

gives us $a_0 = 0.145$, $a_1 = 1.60 \times 10^{-3}$, $a_2 = -2.78 \times 10^{-5}$, and $a_3 = 3.03 \times 10^{-7}$. And the derivative

$$(\partial\rho/\partial P)_T = a_1 + 2a_2 P + 3a_3 P^2 \tag{12}$$

is then obtained. The calculated pressure dependence of the central peak is compared with the experimental data in Fig. 8 by the solid curve. We regard the agreement as rather satisfactory and use it as the reference.

5. Fine Structures

As expressed in Eq. (2), the Raman spectrum of superfluid ^{4}He is determined mainly by two factors; the polarizability tensor and the four-body correlation function of density fluctuations. As already discussed, the polarizability tensor contributes to gradual change of the spectrum.In superfluid helium, density fluctuations are expanded in terms of quasiparticle operators. Therefore, corresponding to the phonon–maxon–roton dispersion relation, the four-body correlation function is expected to show a characteristic form that may be observed as structures in the Raman spectrum.

In Fig. 9, Raman spectra of superfluid ^{4}He measured at a temperature of 1.3 K and at pressures of (a) svp and (b) 20 kg/cm^2 are shown. The instrumental full width at half maximum was 1 K. In the spectra, we confirmed a few fine structures. In the figure, the peak p_1 is the two–roton peak. In addition to it, we can observe a few other peaks, which are labeled by p_2 $- p_5$, and also change of slope $dI_R(\omega)/d\omega$ at a few energy shift values,which are labeled by b – f. Characteristic energy values

$$l\Delta_0 + m\Delta_1 \qquad (\ l,m = \text{positive integers}) \tag{13}$$

are indicated with vertical bars from smaller values up to $4\Delta_1$. Δ_0 and Δ_1 are respectively the minimum of single roton energy and the maximum of single maxon energy determined by the neutron scattering experiments [35,54].

If we assume a linear relation between density and quasiparticle operators as given by Eq. (5), the Raman process corresponds to creation of two elementary excitations having momenta equal in magnitude but directed oppositely. Since the first theoretical proposal by HALLEY[14],only this process has been considered as main possible Raman process. The theory of two-phonon Raman scattering was developed by taking into account interactions between the created two excitations by IWAMOTO [16] and RUVALDS–ZAWADOWSKI [17]. According to these theories, distinct change of spectrum is expected only at $2\Delta_0$ and $2\Delta_1$. As shown in Fig. 9, experimental spectrum shows anomalies not only at these energies but also at other energies. The result suggests existence of other processes also. By numerical caluculation using the model proposed by ZAWADOWSKI, RUVALDS, and SOLANA [40], HASTINGS and HALLEY [20] pointed out a possibility of a four–roton Raman scattering. In Fig. 9(a), the peak p_2 appears at near $4\Delta_0$. It may be assigned as the four-roton peak. Quite recently, HIRASHIMA and IWAMOTO[28] proposed three–phonon Raman processes to explain our experimetal results. IWAMOTO explains detail of the theory in this volume.The theory predicts anomalies just at energies $3\Delta_0$, $2\Delta_0 + \Delta_1$, $\Delta_0 + 2\Delta_1$, and $3\Delta_1$ in addition to $2\Delta_0$ and $2\Delta_1$. Each anomaly is characterized by an infinite slope at these energies. It may be observed as a peak, a cusp, a dip, or an inflection.But it is not certain in which form each anomaly appears. The anomalies c, d, and e, observed in the svp spectrum, closely correspond to characteristic energies $3\Delta_0$, $2\Delta_1$, and $2\Delta_0 + \Delta_1$, respectivly.The peak p_2 is asymmetric and the shoulder on the higher energy side is observed as a weakly resolved peak at slightly higher pressures between 3 and 6 kg/cm^2. (The results are not shown here and will be reported elsewhere soon.) The energy of the shoulder is close to $\Delta_0 + 2\Delta_1$. The very weak anomaly f is not far from $3\Delta_1$. Rather reasonable correspondence of the structures to the multi–phonon characteristic energies supports to assign them to multi–phonon processes. If we are only concerned with the kinematics, because of very small momentum trasfer of light scattering compared with the momentum of individual quasiparticle, multi–phonon processes with zero total momentum are possible. Then we can imagine many possible multi–

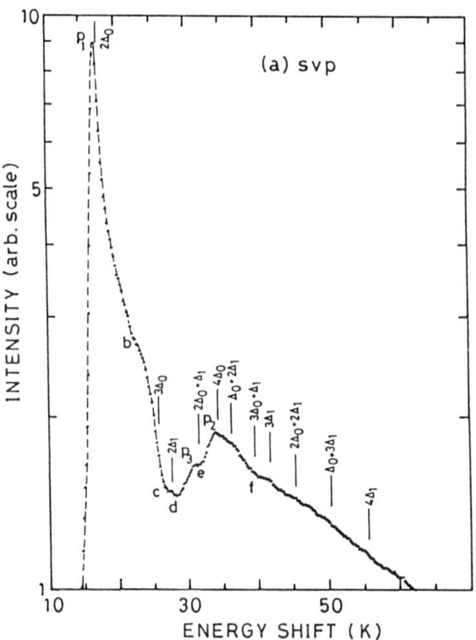

Fig. 9 (a). Raman spectrum of superfluid ⁴He measured at a temperature of 1.3 K and at svp. The instrumental FWHM was 1 K. Three peaks p_1, p_2, and p_3 are observed. We also note change of the slope at a few energies labeled by b – f. Characteristic energy values, $l\Delta_0 + m\Delta_1$ (l, m = positive integers), are shown. The broken line through the data points is hand drawing for the aid of eye.

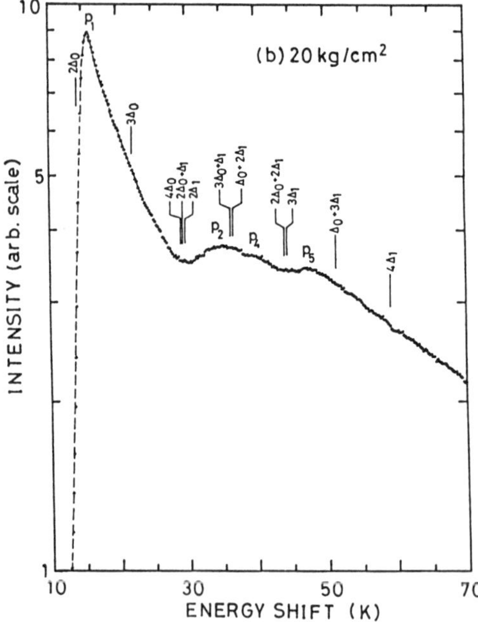

Fig. 9 (b). Raman spectrum of superfluid ⁴He measured at a temperature of 1.3 K and at applied pressure of 20 kg/cm². The instrumental FWHM was 1 K. Four peaks p_1, p_2, p_4, and p_5 are observed. Characteristic energy values, $l\Delta_0 + m\Delta_1$ (l, m = positive integers), are shown. The broken line through the data points is hand drawing for the aid of eye.

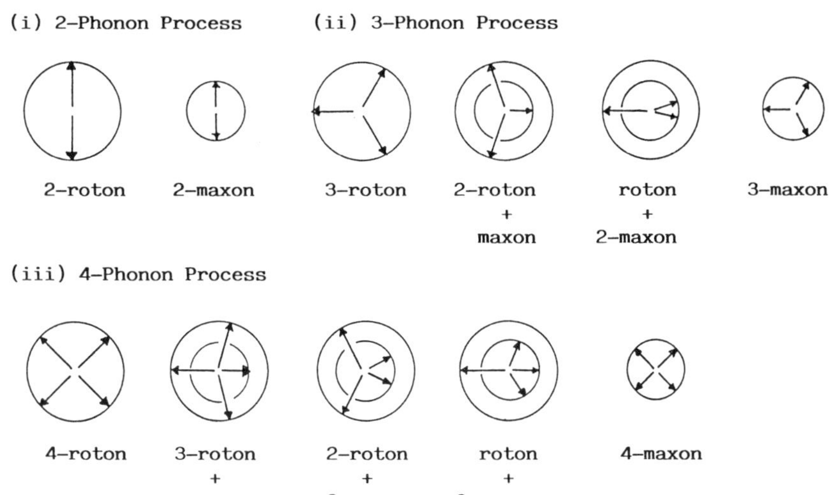

Fig. 10. Wavevector relations of multi-phonon Raman processes. Because the momentum transfer of light scattering is much smaller than the momenta of rotons and maxons, multi-phonon processes with zero total momentum are possible. In the figure, the longer arrows indicate rotons and the shorter arrows maxons.

phonon Raman scatterings by rotons and maxons as shown in Fig. 10. Higher order Raman processes might be possible, which are not shown in the figure.

Contrary to the good correspondence of characteristic energies to those energies at which the fine structures are observed in the svp data, the spectrum observed at 20 kg/cm² does not show good fit as shown in Fig. 9(b) In the spectrum, four peaks p_1, p_2, p_4, and p_5 are observed. From the measurement of pressure dependence, p_2 at this pressure may be regarded to correspond to p_2 in the svp spectrum. However, if we assign it as the four-roton peak, it is shifted to higher energy by about 5 K from $4\Delta_0$. The peaks p_2, p_4, and p_5 are broad and we cannot make clear correspondence of these peaks to multi-phonon characteristic energies. The broadness is not due to temperature broadening but due to intrinsic nature because we observe similar broad peaks at the lower temperature of 0.65K. Because these peaks are observed at energies higher than both $2\Delta_0$ and $2\Delta_1$, it is appropriate to regard them as the peaks corresponding to Raman processes associated with more than two quasiparticles. However, because the experimental feature is quite different from that expected from the theory by Hirashima and Iwamoto, we need to seek some other additional multi-phonon process.

HALLEY discusses in this volume two possible sources of multiparticle processes ;(1) quasiparticle anharmonicities and (ii) the nonlinearities in the relation between the density ρ and the quasiparticle operators α_k. The source of the multiphonon Raman processes in the theories of Iwamoto and Hastings-Halley is interaction between elementary excitations and is quasi-quasiparticle anharmonicity according to Halley's classification. In this case, multi-particle states are mixed with a single particle state via interactions between quasiparticles. And creation of a quasiparticle pair by light necessarily includes higher order multi-particle states. However,non-

linearity in the relation between density and quasiparticle operators also
may be a source of higher order multi-phonon Raman scatterings. In order to
relate dynamic quasiparticle correlations to the Raman cross section given
by Eq.(2), we need the relation beween density and quasiparticle operators.
If it is not the linear relation given by Eq.(5) but includes higher order
nonlinear terms, multiquasiparticle processes appear without interactions
between them. HALLEY shows density fluctuations associated with rotons followed by backflows (short range correlations of atoms near a quasiparticle)
include such nonlinear terms. Importance of the nonlinearity was pointed
out by NAKAJIMA [45] in the early theoretical trials to formulate the Raman
process in superfluid helium.

At present, there is no Raman scattering theory which explicitly takes
into account the nonlinearity. HALLEY suggests calculation of Raman scattering cross section along the line of the theory by MANOUSAKIS and PANDHA-
RIPANDE [41] who calculated $S(k,\omega)$. HALLEY points out that their calculation is free from violation of excluded volume conditions [22,23]. Their
theory yields a theoretical phonon-roton dispersion curve which agrees well
with the experimental result. The theory shows that not only roton-roton
but also roton-maxon and maxon-maxon interactions are important, as the result of which appearance of a few fine structures in the multiphonon region of $S(k,\omega)$ is predicted. Their theoretical results qualitatively resem-
ble experimental observation by STIRLING [29]. Both in the theory and the
experiment, multiphonon peaks do not appear just at the characteristic en-
nergies given by Eq. (13). They are shifted to higher energies.

Rather different feature between the spectrum at svp and that at 20 kg/
cm^2 suggests that the two possible sources work in comparable order and the
main contribution shifts from one to the other as pressure increases. At
lower pressures, we see, the effect of the anharmonicity is dominant. In
this case, anomalies are expected near the characteristic energies as predicted by IWAMOTO. This explains rather good agreement of the svp spectrum
with his theory. As the pressure increases, however, the nonlinear effect
becomes increasingly dominant. This interpretation is consistent with the
experimental fact that qualitative feature of the fine structures sensitively changes with a small change of applied pressure. As the result of this
interpretation, we come to regard that the multiphonon spectrum, due mainly
to nonlinearity, has tendency to give a broad peak positively shifted from
the corresponding characteristic energy. Although there is no theoretical
verification of the interpretation now, future work by theorists will give
us a more refined picture of the multi-phonon processes.

6. Temperature Dependence

Concerning temperature dependence of $S(k,\omega)$ measured by neutron scattering, there are a few controversial problems. Among these, we discuss two
problems here that can be studied by Raman scattering also: One is temperature dependence of the roton energy $\Delta_0(T)$ at fixed pressure and the other
is a two-fluid model of $S(k,\omega)$

DIETRICH ET AL. [54] measured temperature dependence of the roton energy
at several pressures and found strong decrease of $\Delta_0(T)$ as the temperature
approaches toward the lambda point. At svp, the decrease of $\Delta_0(T)$ was about
70 %. In order to determine the roton energy at finite temperature, they
assumed a lorentzian function of the neutron scattering peak by rotons.
However, later on, TARVIN and PASSEL [30] found that a better fit of the

data is observed if a harmonic oscillator model is used instead of a lorentzian function. The model yields higher roton energies at higher temperatures where considerable broadening of the neutron scattering roton peak is
observed. At low temperatures, where the broadening is not significant, the
two models give the same result. They found little temperature dependence
of the roton energy $\Delta_0(T)$ if the experimental data were analyzed with the
harmonic oscillator model. At svp, decrease of the roton energy was less
then 6 %.

In the case of Raman scattering, we can observe the two-roton peak, position of which tells us the energy of interacting two rotons. We measured
temperature dependence of the Raman scattering at several pressures and
found that the two-roton energy shows little temperature dependence. An example, measured under the applied constant pressure of 1 kg/cm^2, is shown
in Fig. 11. In the figure, two spectra measured at temperatures of 1.3 K
and 1.91 K are shown in (a) and (b), respectively. Between the two temperatures, practically no shift of the two-roton peak is detected. The energy
$2 \Delta_0$, equal to twice the energy of a single roton estimated by DIETRICH ET
AL. [54] from their neutron experiment using the lorentzian function, is
shown. The energy is considerably smaller than the position of the two-
roton Raman peak at 1.91 K. TARVIN and PASSEL [30] reported the result only
at svp. Our Raman measurements shown in Fig. 11 were done at 1 kg/cm^2. Because of the small difference of the pressures, we can expect similar neutron result at 1 kg/cm^2 also; i.e. a very small change of the roton energy
$\Delta_0(T)$ with temperature. Thus our result supports the analysis by TARVIN and
PASSEL using the harmonic oscillator model to fit the experimental neutron
scattering spectrum, which gives very weak temperature dependence of the
roton energy.

Concerning the temperature dependence of $S(k,\omega)$, WOODS and SVENSSON [36]
proposed a two-fluid model (the Woods-Svensson model). They proposed that
$S(k,\omega)$ is expressed as a sum of two terms, one is proportinal to superfluid
density ρ_s and the other is proportional to normal fluid density ρ_n. ISHI-
KAWA and YAMADA discuss the model in this volume theoretically.   According
to the Woods-Svensson model, the superfluid part of $S(k,\omega)$ decreases as
temperature increases toward the lambda point and disappears above it. The
normal part of $S(k,\omega)$ increases with an increase of temperature toward the
lambda point and becomes nearly constant above it. Thus the sum of the two
parts, i.e. the observed $S(k,\omega)$, is expected to show a strong temperature
dependence below the lambda point and becomes almost constant above it. In
fact WOODS and SVENSSON [36] observed such behavior of $S(k,\omega)$ at svp and at
a wavevector of 1.13 $\overset{\circ}{A}^{-1}$. They observed a good fit of the experimental temperature dependence of $S(k,\omega)$ to their model. Although the Raman spectrum
is not directly proportional to $S(k,\omega)$, close relations exist between them.
And, if the two-fluid model is really applicable to $S(k,\omega)$, then we can observe related effect of it on the Raman spectrum. For example, we might observe a strong decrease of the intensity toward the lambda point and almost
constant value above it. In the inset of Fig. 11(a), temperature dependence
of the two-roton peak intensity is shown. We observed a smooth change of
the intensity of the two-roton peak even near the lambda point. The intensity is not constant above the lambda point but decreases with an increase
of temperature. This fact is not consistent with the Woods-Svensson model,
which claims a nearly constant value above the lambda point. However, there
is a feature that is favorable to the model. In Fig. 11 (b), the broken
curve is the result of the convolution integral of the lower temperature
spectrum Fig. 11 (a) with a lorentzian function to take into account a tem-
peature broadening effect. The width of the lorentzian was chosen so that
the lower energy side of the two-roton spectrum nearly traces the experi-

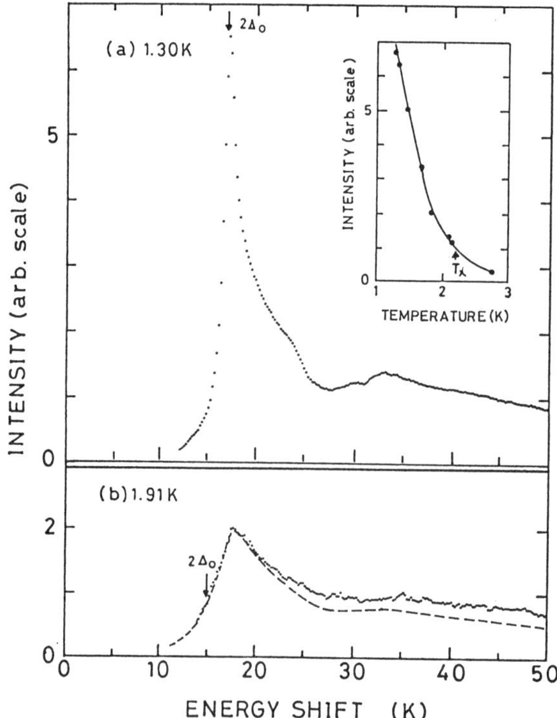

Fig. 11. Raman spectra of superfluid ⁴He measured at (a) 1.30 K and (b) 1.91 K, at an applied pressure of 1 kg/cm². The instrumental FWHM was 1 K. $2\Delta_0$ is twice the energy of a single roton measured by neutron scattering. In (b), the broken curve is the result of the convolution integral of the spectrum shown in (a) with a lorentzian function, whose width was chosen so that the lower energy side of the two-roton peak fits well. Inset: Temperature dependence of the two-roton peak intensity. T_λ is the lambda point.

mental data. As shown in the figure, the fit at higher energy side of the peak is not so good.This discrepancy is eliminated if we add an appropriate portion of the normal state Raman spectrum, in the present case the Raman spectrum measured at 2.73 K. The calculated spectrum closely traces the experimental shape shown in Fig. 11 (b). (The result is not shown here.) This fact suggests that the two-fluid model, at least qualitatively, gives a better explanation of the temperature dependence of the Raman spectrum than to consider the temperature effect only by the lorentzian broadening of the spectrum. However, in the case of Raman spectrum, the ratio of the the normal part to the superfluid part does not quantitatively agree with the observed ratio of the normal density to the superfluid density at the temperature.

Acknowledgements

The author wishes to express his thanks to many people, in particular, J. W. Halley, F. Iwamoto, S. Nakajima , C. E. Campbell, and E. C. Svensson for their stimulating discussions. The author also wishes to thank to his colleagues, **M.** Watabe, **M.** Udagawa, H. Nakamura, **M.** Murakami, T. Imada, T. Akagi, **N.** Ogita, Y. Utsunomiya, and A. Fukumoto for their enthusiastic collaborations.

References

1. K. Ohbayashi and M. Udagawa, Phys. Rev. B 31, 1324(1985).
2. M. Udagawa, H. Nakamura, M. Murakami, and K. Ohbayashi, Phys. Rev. B 34, 1563(1986).
3. M. Udagawa, T. Imada, and K. Ohbayashi, J. Phys. C 20, 1063(1987).
4. K. Ohbayashi, M. Udagawa, and M. Watabe, Can. J. Phys. 65, 1571(1987).
5. K. Ohbayashi, T. Akagi, N. Ogita, M. Watabe and M. Udagawa, Jpn. J. Appl. Phys. 26, Supplement 26-3, 5(1987).
6. M. Udagawa, Y. Utsunomiya, A. Fukumoto, M. Watabe, and K. Ohbayashi, Jpn. J. Appl. Phys. 26, Supplement 26-3, 35(1987).
7. L.D. Landau and E.M. Lifshitz, Electrodynamics of Continuous Media (Addison-Wesley, Mass, 1960), Chap. XIV.
8. W.F. Vinen, In Physics of Quantum Fluids, eds. R. Kubo and F. Takano (Shokabo, Tokyo 1971) p.1.
9. T.J. Greytak, In Quantum Liquids, ed. by J. Ruvalds and T. Regge (North-Holland, Amsterdam. 1978) p.121.
10. M.J. Stephen, In The Physics of Liquid and Solid Helium, ed. by K.H. Bennemann and J.B. Ketterson (Wiley, New York, 1976), Part 1, Chap. 4.
11. J.W. Halley (ed.), Correlation Functions and Quasiparticle Interactions in Condensed Matter, (Plenum Press, New York, 1978).
12. K. Ohbayashi and A. Ikushima, J. Low Temp. Phys. 15, 33(1974).
13. J.A. Tarvin, F. Vidal, and T.J. Greytak, Phys. Rev. B 15, 4193(1977).
14. J.W. Halley, Phys. Rev. 181, 338(1969).
15. T.J. Greytak and J. Yan, Phys. Rev. Lett. 22, 987(1969).
16. F. Iwamoto, Prog. Theor. Phys. (Japan), 44, 1135(1970).
17. J. Ruvalds and A. Zawadowski, Phys. Rev. Lett. 25, 333(1970).
18. T.J. Greytak, R.L. Woerner, J. Yan, and R. Benjamin, Phys. Rev. Lett. 25, 1547(1970).
19. C.A. Murray, R.L. Woerner, and T.J. Greytak, J. Phys. C 8, L90(1975).
20. R. Hastings and J.W. Halley, Phys. Rev. A 10, 2488(1974).
21. J.W. Halley and R. Hastings, Phys. Rev. B 15, 1404(1977).
22. P. Kleban, Phys. Lett. 49A, 19(1974); Phys. Rev. B 19, 3511(1979).
23. P. Kleban and R. Hastings, Phys. Rev.B 11, 1878(1975).
24. P. Kleban and J. W. Halley, Phys. Rev. B 11, 3520(1975).
25. F.J. Pinski and C.E. Campbell, Chem. Phys. Lett. 56, 156(1978).
26. C.E. Campbell, J.W. Halley, and F.J. Pinski, Phys. Rev. B 21, 1323 (1980).
27. K. Ohbayashi and A. Ikushima, J. Phys. C 7, L206(1974).
28. D.S. Hirashima and F. Iwamoto, Prog. Theor. Phys.(Japan) 75, 744(1986).
29. W.G. Stirling, In Proc. Second Intl. Conf. on Phonon Physics, eds. J. Koller, N. Kroo, N. Menyhard, and T. Siklos (World Scientific, Singapore, 1985).
30. J.A. Tarvin and L. Passel, Phys. Rev. B 19, 1458(1979).
31. A.D.B. Woods and E.C. Svensson, Phys. Rev. Lett. 41, 974(1978).
32. J. Ruvalds and T. Regge (eds.), Quantum Liquids (North-Holland, Amsterdam, 1978).
33. R.A. Cowley and A.D.B. Woods, Can. J. Phys. 49, 177(1971).
34. L.P. Pitaevskii, Sov. Phys. JETP 9, 830(1959).

35. E.H. Graf, V.J. Minkiewicz, H.B. Moller, and L. Passell, Phys. Rev. A 10, 1748(1974).
36. A.D.B. Woods, E.C. Svensson, P. Martel, Phys. Lett. A 43, 223(1973).
37. E.C. Svensson, P. Martel, V. F. Sears, and A.D.B. Woods, Can. J. Phys. 54, 2178(1976).
38. A.J. Smith, R.A. Cowley. A.D.B. Woods, W.G. Stirling, and P. Martel, J. Phys. C 10, 543(1977).
39. K. Nagai, J. Phys. C 11, L759(1978).
40. A. Zawadowski, J. Ruvalds, and J. Solana, Phys. Rev. A 5, 399(1972).
41. E. Manousakis and V.R. Pandharipande, Phys. Rev. B 33, 150(1986).
42. K.S. Bedel, D. Pines, and A. Zawadowski, Phys. Rev. B 29, 102(1984).
43. M.J. Stephen, Phys. Rev. 187, 279(1969).
44. A.L. Fetter, J. Low Temp. Phys. 6, 487(1972).
45. S. Nakajima, Prog. Theor. Phys. (Japan) 45, 353(1971).
46. R.L. Woerner and T. J. Greytak, J. Low Temp. Phys. 13, 149(1973).
47. L. Frommhold, Adv. Chem. Phys. 46, 1(1981).
48. P.D. Dacre and L. Frommhold, J. Chem. Phys. 76, 3447(1982).
49. R.A. Aziz, V.P.S. Nain, W.L. Taylor, and G.T. MacConville, J. Chem. Phys. 70, 4330(1979).
50. P.C. Hohenberg and P.M. Platzman, Phys. Rev. 152, 196(1966).
51. D. Baeriswyl, Phys. Lett. 41A, 297(1972).
52. D. Baeriswyl, Geneva thesis (No.1628, Dept. Theoret. Phys.,1974).
53. J. Wilks, The Properties of Liquid and Solid Helium, (Oxford University Press, 1967).
54. O.W. Dietrich, E.H. Graf, C.H. Huang, and L. Passell, Phys. Rev. A 5, 1377(1972).

Polarization Study of Raman Scattering from Quantum Fluids

Masayuki Udagawa, Mitsuo Watabe, and Kohji Ohbayashi

Faculty of Integrated Arts and Sciences, Hiroshima University,
Hiroshima 730, Japan

Depolarization properties of Raman spectra of simple liquids have been studied experimentally by decomposing them into polarized (s-wave) and depolarized (d-wave) components. For quantum liquids, ^{3}He and ^{4}He, the ratio of the polarized component to the total intensity has been observed to increase in the large energy shift region. Pressure dependence has been measured both for super and normal fluid ^{4}He. For liquid ^{4}He, the integrated intensity ratios of the s-wave component to the d-wave component have been estimated and found to increase with increasing pressure. The values of the ratios for ^{4}He are consistent with the theoretical values by Campbell et al. For Ne, we have observed weak polarized component in the large energy shift region, but the contribution of the polarized intensity to the total intensity is smaller compared with those of quantum liquids. For Ar, the ratio of polarized component to the total intensity is even weaker than that of Ne. Existence of the polarized component suggests that the repulsive interaction is also important to investigate the Raman scattering process of simple liquids as well as conventional dipole-induced-dipole interaction. Relative contribution of the polarized component to the total intensity has been found to decrease as a liquid loses quantum nature.

1. INTRODUCTION

Raman scattering is a useful tool to investigate dynamical properties of simple liquids. Raman scattering rate $I_R(\omega)$ of simple fluids may be given in a simplified expression as

$$I_R(\omega) \propto \int dk^3 \int dk'^3 \, t(k) t(k') H(k,k',\omega), \tag{1}$$

where $t(k)$ is a light-atom coupling function and $H(k,k',\omega)$ is a Fourier-transformed density correlation function. $H(k,k',\omega)$ mainly determines Raman spectral shape and gives informations of dynamical atomic arrangement. $t(k)$ determines the polarization properties and is closely related to inter-atomic interaction. We can imagine two different limits of the interatomic separation. For a large inter-atomic separation, a dipole-induced-dipole (DID) interaction due to van der Waals force is important. While for a short interatomic separation, a repulsive interaction due to an electron overlap effect (EO) may be important. Depolarization property of Raman scattered light depends on these types of the interatomic interaction : DID gives rise to only depolarized d-wave scattering [1], while EO gives both polarized s-wave and depolarized d-wave scattering [2]. The zero point vibration, i.e. quantum nature, is expected to affect the short-distanced EO interatomic interaction. Thus the polarized s-wave contribution, due to the EO interaction, is expected to change depending on degree of quantum nature of a fluid. In order to see this relationship, we have studied the polarization properties of a series of liquids ^{3}He, ^{4}He, Ne, and Ar.

Springer Series in Solid-State Sciences, Vol. 79   **Elementary Excitations in Quantum Fluids**
Editors: K. Ohbayashi · M. Watabe   © Springer-Verlag Berlin, Heidelberg 1989

2. EXPERIMENTAL

A 514.5nm radiation of an Ar ion laser was used as the incident light. The scattered light was collected with a F3.7 lens at 90° to the incident beam direction and was analyzed by a double grating spectrometer (Jobin-Yvon model HG2S). Output light was detected by a photomultiplier tube (Hamamatsu Photonics model No. R464) and was amplified-discriminated by a photon counter (Princeton Applied Research model No.1121).

In order to separate two differently polarized components, we employed the following two different ways of observation as shown in Fig.1 ; I_{vv} and I_{hv}. In I_{vv}, the incident beam was vertically polarized and the easy axis of the polarizer of the scattered light was vertical. In I_{hv}, the polarization direction of the incident beam was horizontal and the easy axis of the analysing polarizer was vertical. To change the direction of the incident beam polarization from vertical to horizontal, we used a 1/2 wavelength plate. Considering the scattered intensity $I_R(\omega)$ is a sum of polarized component $I_p(\omega)$ and depolarized component $I_{dp}(\omega)$,

$$I_R(\omega) = I_p(\omega)(\hat{e}_i \cdot \hat{e}_s)^2 + I_{dp}(\omega), \tag{2}$$

where $\hat{e}_i$ and $\hat{e}_s$ denote the polarization unit vectors of the incident and scattered light, respectively. For the present experimental conditions, eq.(2) gives

$$I_{vv}(\omega) = I_p(\omega) + I_{dp}(\omega) \qquad\qquad I_{hv}(\omega) = I_{dp}(\omega). \tag{3}$$

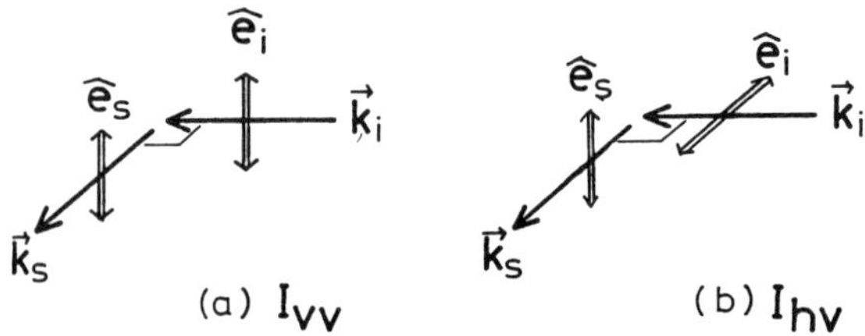

Fig.1 Polarization conditions. (a) and (b) correspond to I_{vv} and I_{hv}, respectively. $\hat{e}_i$ is unit vector parallel to the incident beam polarization, $\hat{e}_s$ unit vector of the easy axis of the analyzing polarizer, $\vec{k}_i$ wavevector of incident beam, and $\vec{k}_s$ wavevector of scattered light.

3. RESULTS AND DISCUSSION

3.1 SUPERFLUID ^{4}He

Obtained Raman spectra of I_{vv} and I_{hv} are shown in Fig.2. Fig.2(a) and Fig.2(b) were taken with a 3.7K full width at half maximum (FWHM) at pressures of 1.0 and 21.1 kg/cm^2, respectively. The structures below the energy shift ~60K are related to elementary excitations of ^{4}He [3,4].

With an increase of the energy shift, the ratio I_{vv}/I_{hv} becomes larger in the large energy shift region. The same tendency was also observed at 6.2 kg/cm^2. According to Iwamoto [5], Raman scattering intensity $I_R(\omega)$ for Bose fluid may be regarded as composed of a s-wave and a d-wave as

$$I_R(\omega) = I_s(\omega)(\hat{e}_i \cdot \hat{e}_s)^2 + I_d(\omega)\{0.25(\hat{e}_i \cdot \hat{e}_s)^2 + 0.75\}, \tag{4}$$

54

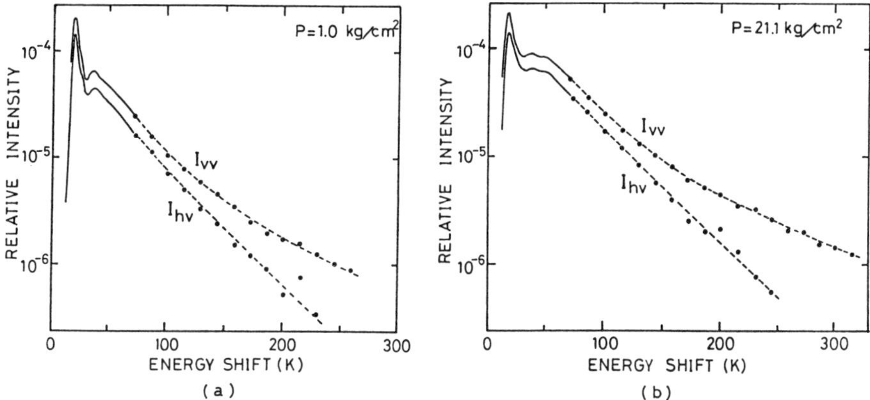

Fig.2 I_{vv} and I_{hv} Raman spectra of superfluid ^{4}He. These spectra were measured at 1.36K with a 3.7K FWHM; (a) 1.0 kg/cm^2, (b) 21.1 kg/cm^2. The vertical axis shows intensity relative to central Brillouin scattering intensity.

where $I_s(\omega)$ and $I_d(\omega)$ denote the scattering intensity of the s-wave and d-wave, respectively. Considering the polarization conditions of our experiments, $I_s(\omega)$ and $I_d(\omega)$ are

$$I_s(\omega) = I_{vv}(\omega) - I_{hv}(\omega)/0.75$$
$$\text{and}\quad I_d(\omega) = I_{hv}(\omega)/0.75. \tag{5}$$

The estimated s-wave and d-wave at 1.0 kg/cm^2 are shown in Fig.3. Also in higher pressure spectra, the contribution of the s-wave part becomes dominant in the large energy shift region.

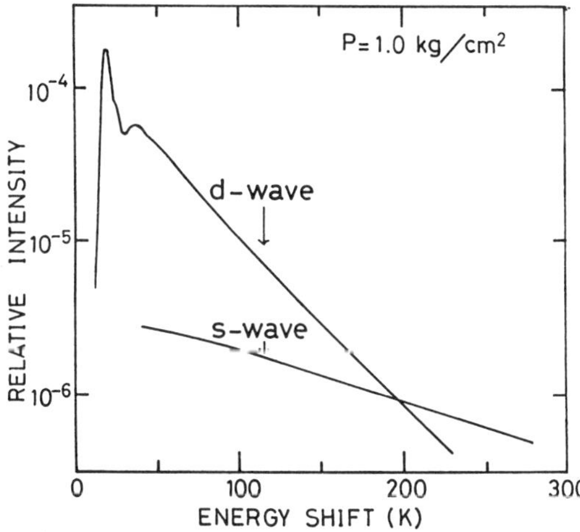

Fig.3 Estimated s- and d-wave scattering intensities. The value of the vertical axis is intensity relative to Brillouin scattering.

Here we compare our experimental results with theories. To make full
comparison of our results, we need theoretical forms of both polarized and
depolarized spectra over all energy shift region. However there has been no
theoretical report of them. An available quantity of the theoretical calculation is a ratio of the integrated intensities I_s/I_d. The experimental
results of the ratio of the integrated intensity I_s/I_d is listed in table 1.
Integrations were done over the energy shift region from 10K to infinity,
assuming the functional form of the intensity in the large energy shift
region to be single exponential. For details, we refer to reference 6.
Kleban and Halley made a semiempirical calculation [7], using the experimental results of the ultraviolet absorption spectra measured by Surko et
al.[8] Their results predicted $0.11 \leq I_s/I_d \leq 1.45$. The uncertainty was
estimated from the approximation made for the correlation function and from
incomplete information of the ultraviolet absorption line shape. Another
numerical calculation was made by Campbell et al.[9]. They related the
integrated intensity ratio to the trace of the pair polarizability tensor
of an interacting diatom. Their results predicted $0.08 \leq I_s/I_d \leq 1.88$.
Their results show that the ratio I_s/I_d increases with an increase of the
contribution of the EO interaction. Present results lie within (but rather
near the smallest end of) the region predicted by the theoretical calculations. The increase of I_s/I_d with an increase of the pressure can be
attributed to the increase of the contribution of the repulsive interaction due to the decrease of interatomic distance.

Table 1 Estimated integrated ratio of s-wave intensity to d-wave intensity.

Temperature (K)	Pressure (kg/cm^2)	I_s/I_d
1.36	1.0	0.10
1.36	6.2	0.16
1.36	21.1	0.14
4.2	2.4	0.12
4.2	6.1	0.11
4.2	10.9	0.12
4.2	16.1	0.13
4.2	21.0	0.14

3.2 NORMAL LIQUID ^{4}He

The obtained spectra at 4.2K under 2.4kg/cm^2 are shown in Fig.4. We can see
clearly a broad peak at ~30K both for I_{vv} and I_{hv}. The energy shift of this
peak increases with an increase of pressure. The intensity ratio I_{vv}/I_{hv}
increases with an increase of the energy shift in the large energy shift
region. We also estimated the contribution of I_s and I_d. The integrated
ratios I_s/I_d are also estimated at different pressures. These results are
tabulated in table 1. The ratio I_s/I_d increases also with an increase of
pressure in normal liquid ^{4}He. These obtained values are also consistent
with the calculated values by Campbell et al.[9]. Then we can conclude that
the repulsive interaction is also important for t(k) in normal fluid ^{4}He.

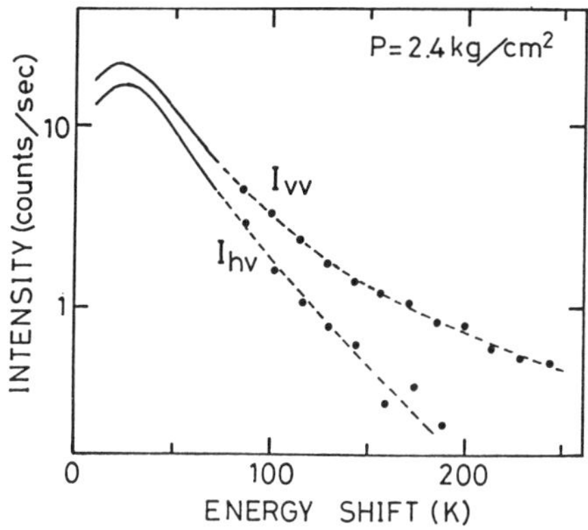

Fig.4 Polarization properties of Raman spectra of normal fluid ^{4}He at 4.2K.
These spectra were measured at 2.4 kg/cm^2 with a 4.3K FWHM.

3.3 NORMAL LIQUID ^{3}He

I_{vv} and I_{hv} Raman spectra of the normal liquid ^{3}He are shown in Fig.5.
These spectra were measured at 1.2K under a pressure of 2.05 kg/cm^2. We
found that the whole spectral shape of normal liquid ^{3}He [10] is very similar to that of normal liquid ^{4}He. The intensity ratio I_{vv}/I_{hv} increases
with an increase of the energy shift in the large energy shift region.

Since liquid ^{3}He is a Fermi liquid, there is a possibility of another symmetry scattering wave such as a p-wave. Then we cannot apply eq.(4) to liquid
^{3}He to obtain only I_s and I_d. In this experiment, however, a difference
of statistics between ^{3}He and ^{4}He could not be seen clearly. To clarify the
difference of the statistics, we need an experiment at a much lower temperature.

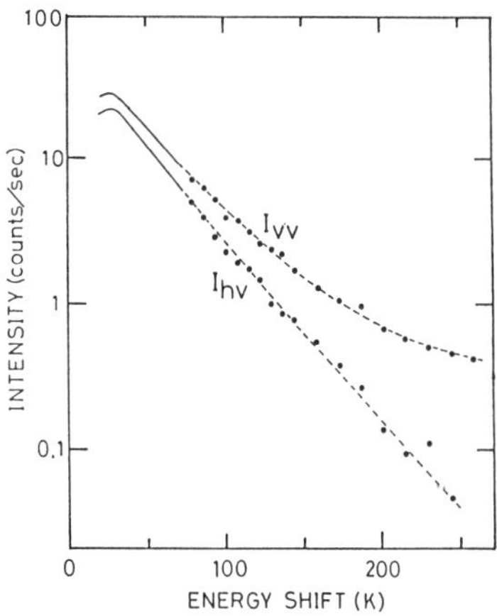

Fig.5 I_{vv} and I_{hv} Raman spectra of liquid ^{3}He. These spectra were measured
at 1.2K and 2.05 kg/cm^2 with a 9.2K FWHM.

3.4 LIQUID Ne

Raman spectra of liquid Ne were measured in the temperature region between
26 and 41K. There is no distinct structure. As the energy shift increased,
we could detect a systematic increase of the intensity ratio I_{vv}/I_{hv}, which
verifies existence of polarized s-wave scattering. The ratio of the polarized light intensity to the total intensity is smaller for liquid Ne, compared with those for liquid ^{3}He and liquid ^{4}He.

3.5 LIQUID Ar

We have measured Raman spectra of liqud Ar in the temperature region between 86 and 112K. I_{vv} and I_{hv} decrease with an increase of the energy
shift with a functional form of nearly single exponential. The intensity
ratio I_{vv}/I_{hv} did not show a clear increase even in the large energy shift
region.

3.6 CONCLUSION

The existence of the polarized component suggests that the EO interaction
is also important as well as the DID interaction for the Raman scattering
process of simple fluids. For ^{4}He, the integrated ratios I_s/I_d were
estimated, and the values and their pressure dependence are consistent
with the predicted values by theories [7,9]. By comparing the relative
contribution of the polarized component to the total intensity for a series
of simple fluids ^{3}He, ^{4}He, Ne, and Ar, it was found to increase with an
increase of quantum nature in these simple liquids.

REFERENCES

1 M. J. Stephen : Phys. Rev. 187 (1969) 279.
2 S. Nakajima : Prog. Theor. Phys. 45 (1971) 353.
3 K. Ohbayashi and A. Ikushima : J. Phys. C7 (1974) L206.
4 K. Ohbayashi and M. Udagawa : Phys. Rev. B31 (1985) 1324.
5 F. Iwamoto : Prog. Theor. Phys. 44 (1970) 1135.
6 M. Udagawa, H. Nakamura, M. Murakami, and K. Ohbayashi : Phys. Rev. B34
   (1986) 1563.
7 P. Kleban and J. W. Halley : Phys. Rev. B11 (1975) 3520.
8 C. M. Surko, G. J. Dick, F. Reif, and W. C. Walker : Phys. Rev. Lett.
   23 (1969) 842.
9 C. E. Campbell, J. W. Halley, and F. J. Pinski : Phys. Rev. B21 (1980)
   1323.
10 M. Udagawa, T. Imada, and K. Ohbayashi : J. Phys. C20 (1987) 1063.

Temperature Dependence of $S(Q, \omega)$ for Liquid ^{4}He

E.C. Svensson

Atomic Energy of Canada Limited, Chalk River Nuclear Laboratories,
Chalk River, Ontario K0J 1J0, Canada

1. Introduction

It was most appropriate to hold this symposium in a year when two very important helium anniversaries were celebrated – the 40th anniversary of LANDAU'S
[1] now famous phonon-roton dispersion relation for the elementary excitations
in superfluid ^{4}He (see Fig. 1), and the 30th anniversary of the first experimental observation, using neutron scattering techniques, of rotons by PALEVSKY
et al. [2]. Their measurements and others [3–6] which quickly followed, all
stimulated by the suggestion of COHEN and FEYNMAN [7] that neutron scattering
could be used to directly determine the helium dispersion relation, confirmed
that the form proposed by Landau was indeed correct.

There have been an enormous number of subsequent neutron scattering studies
of the dynamics of liquid ^{4}He, the results of which have been extensively discussed in numerous review articles [8–13]. The phonon-roton dispersion
relation for liquid ^{4}He at saturated vapor pressure (SVP) and low temperature
(T < 1.2 K) is without doubt the most extensively studied and most accurately
known of all dispersion relations. In Fig. 1 we show the results of a compilation of experimental values by DONNELLY et al. [14]. Note the especially

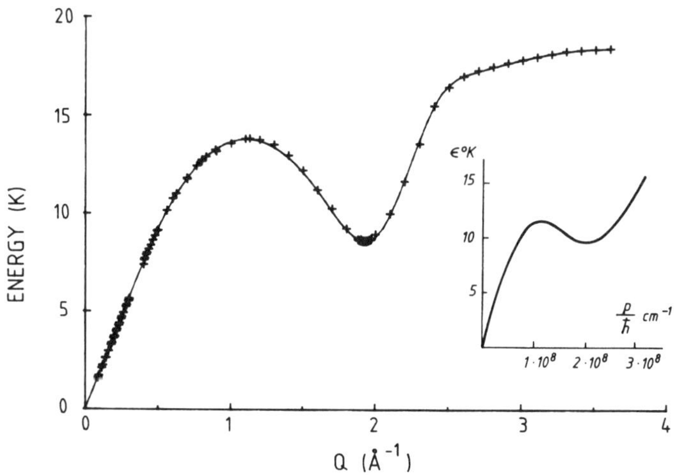

Fig. 1. The one-phonon dispersion relation for liquid ^{4}He at SVP and T < 1.2 K
from a compilation of results by DONNELLY et al. [14]. The inset shows the
dispersion relation proposed by LANDAU [1] in 1947

high density of points at low wave vectors, Q, and in the region of the roton
minimum (Q ≈ 2 $Å^{-1}$). Note also that Landau's prediction (inset), based on the
rather meagre thermodynamic data available in 1947, was extremely close as
regards the position of the roton minimum (1.95 $Å^{-1}$ compared with 1.93 $Å^{-1}$
from the neutron experiments [14]) and only about 10% high as regards the
roton energy gap, Δ.

In spite of the prodigious wealth of experimental results that have been
accumulated over the past three decades, and a large amount of theoretical
work as well, there are still a number of serious gaps in our understanding of
liquid ^{4}He. Most importantly, we do not yet have a detailed understanding of
how the presence of the Bose condensate in the superfluid phase affects the
static and dynamic properties, and it is essential that we do so if we are to
claim a true fundamental understanding of liquid ^{4}He. The experimental
studies that can best guide us to such an understanding are ones in which we
systematically vary the condensate fraction, n_0, for example by varying the
temperature and/or the pressure, and observe what happens to the dynamic
structure factor, $S(Q,\omega)$, and the static structure factor, $S(Q) = \int S(Q,\omega)d\omega$.
(Here $\hbar Q$ and $\hbar\omega = h\nu$ are, respectively, the momentum and energy transferred to
the sample in the scattering process.)

In this paper, I will concentrate primarily on the temperature dependence
of $S(Q,\omega)$. I will further restrict myself largely to a discussion of the
results from neutron scattering studies. To set the stage, I will first make
some general remarks about neutron scattering measurements and techniques and
review some general features of $S(Q,\omega)$ and the dispersion relation at low
temperatures. I will then discuss the temperature dependence of $S(Q,\omega)$ at SVP
and the two-fluid model for $S(Q,\omega)$ proposed by WOODS and SVENSSON [15].
Following this, I will discuss the temperature dependence of $S(Q,\omega)$ at high
pressure and then end with a summary and some remarks about future needs and
directions. The temperature dependence of $S(Q,\omega)$ for liquid ^{4}He, especially
the temperature dependence of the widths and frequencies of the single-roton
excitations, has now been of interest for 3-4 decades. This topic has
generated, and continues to generate, a great deal of discussion, disagreement, uncertainty and confusion. Many of the problems are attributable to a
widespread failure to appreciate the reasons for the differences between
different sets of results and to distinguish between those facts and features
that are firmly established by experiment and those that are the result of the
analysis in terms of some particular model. In the hope of clarifying the
situation, I intend to be particularly thorough in this paper, and sometimes
deliberately repetitive. Since there are still plenty of gaps in our knowledge and understanding, I don't expect, or even want, to eliminate all the
controversy, but I will try to make clear to the reader what I believe, or at
least favor, at the present time. Results from several sources will be
presented in this paper and, since the fad in units has changed over the many
years of the studies on liquid ^{4}He, it may be useful for the reader to note
that 1 THz is equivalent to 48.0 K or 4.14 meV and that 1 bar is equivalent to
0.987 atm or 0.1 MPa.

2. Neutron Scattering and $S(Q,\omega)$

During the discussions at the Hiroshima Symposium, it became clear that the
reasons for the disagreements between the results (especially those for the
temperature dependence of the one-phonon frequencies and widths) from different neutron scattering studies were not in general very well appreciated.
Researchers were thus often uncertain as to which, if any, of the results were
reliable. Although, as we shall see later, the disagreements are often in
large part attributable to differences in the analysis procedures employed,

differences in the "raw" results caused by the different ways in which the measurements were carried out are also an important factor. To help the reader appreciate the latter, I include here a few remarks on techniques.

It is possible to carry out a neutron scattering experiment in such a way that, after relatively straightforward corrections for such effects as fast-neutron background, scattering by the specimen container, order contamination in the neutron beam and multiple scattering, the observed intensity is directly proportional to $S(Q,\omega)$, broadened of course by the experimental resolution. This is the case if one uses a triple-axis crystal spectrometer [16] with a fixed scattered-neutron energy and counts, at each point of the scan (i.e., each ω value), until a given number of neutrons are detected in an incident-beam monitor having a sensitivity inversely proportional to the neutron velocity (e.g., a thin fission detector). A triple-axis crystal spectrometer has the further advantage that one can scan over $S(Q,\omega)$ in ω while keeping Q fixed (Constant-Q technique [16]). For other modes of operation of a triple-axis crystal spectrometer (e.g., with a fixed incident-neutron energy) or for other types of spectrometer (e.g., a time-of-flight spectrometer), the observed intensity may be a very poor approximation to $S(Q,\omega)$, the most common distortion being a severe depression of the high-frequency part.

While the precise way in which the measurements are carried out may be of little importance if $S(Q,\omega)$ consists of only sharp peaks (e.g., as in a harmonic solid), it is very important if $S(Q,\omega)$ extends over a broad frequency range as it does for liquid ^{4}He (see Fig. 2). Relatively few of the vast number of neutron scattering measurements on liquid ^{4}He have been carried out in such a way as to give results which are a close approximaton to the full resolution broadened $S(Q,\omega)$. Of those studies that are of primary interest for this article, namely studies covering the temperature range from low temperatures to T_λ and above, the only ones to give such results are those of WOODS and SVENSSON [15] and SVENSSON et al. [17] at SVP and those of SVENSSON et al. [18] and TALBOT et al. [19] at a pressure of 20 bars.

3. $S(Q,\omega)$ at Low Temperature

Some examples of resolution-broadened dynamic structure factors for superfluid ^{4}He obtained [20] from neutron scattering measurements at low temperature (1.2 K) are shown in Fig. 2. We see that each distribution consists of a sharp low frequency peak which corresponds to a one-phonon excitation on the phonon-roton dispersion relation (see Fig. 1) and, primarily at higher frequencies, a broad multiphonon component which exhibits considerable structure especially for the maxon wave vector ($Q = 1.13$ $\text{\AA}^{-1}$). Except at the higher pressure for the maxon position where there is significant intrinsic broadening, the one-phonon peaks in Fig. 2 have essentially the resolution width. Note that the intensity of the one-phonon peak decreases by more than a factor of two between low and high pressures for the maxon. For $Q = 2.05$ $\text{\AA}^{-1}$, i.e. in the roton region, there is, however, a slight increase in the one-phonon intensity between low and high pressures. This is largely accounted for by the lower frequency at the higher pressure.

The results of the very extensive study of COWLEY and WOODS [21] at SVP and 1.1 K are summarized in Fig. 3. The one-phonon dispersion relation had, by this time, such a high density of points (see Fig. 7 of [21]) that it is shown simply as a solid curve. Also shown in Fig. 3 are the peak energies and upper and lower half-height positions of the broad multiphonon component. (The points for $Q = 0.3$ $\text{\AA}^{-1}$ are the half-height positions from the later measurements of [20].) At low Q, the multiphonon component is a rather narrow peak centred near but significantly above 2Δ. Its peak position then passes

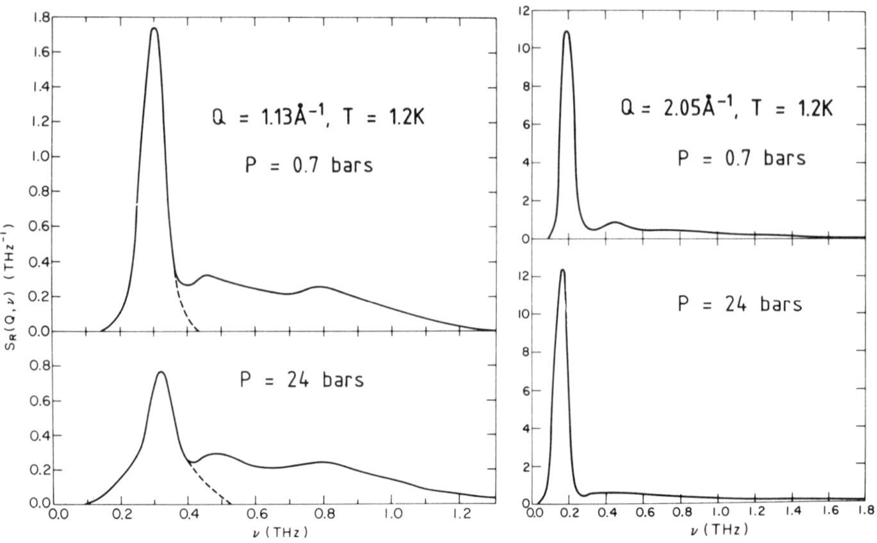

Fig. 2. Resolution-broadened dynamic structure factors for liquid ^{4}He at low and high pressures at 1.2 K. The dashed curves show a plausible division into one-phonon and multiphonon components. From SVENSSON et al. [20]

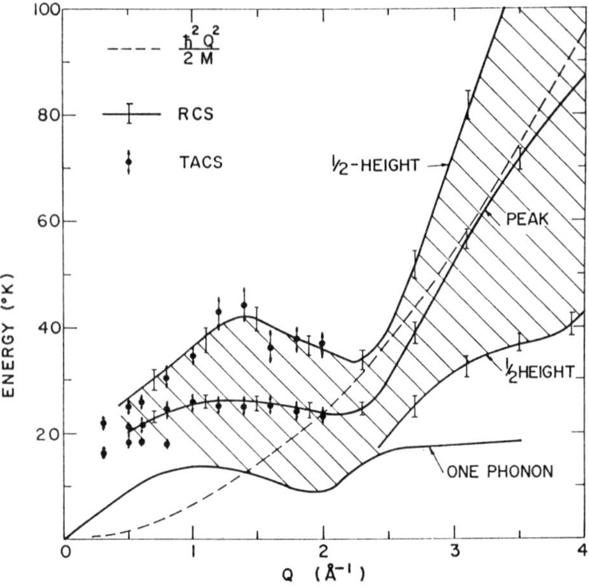

Fig. 3. The one-phonon dispersion relation and the upper and lower half heights and mean energy of the multiphonon peak for liquid ^{4}He at SVP and T = 1.1 K. Adapted from COWLEY and WOODS [21]

through a broad maximum before rising (beyond $Q \approx 2.4$ Å^{-1}) to become centred
near the dispersion relation for free ^{4}He atoms (dashed curve). Its width
also passes through a maximum before increasing rapidly beyond $Q \approx 2.5$ Å^{-1} and
in fact continues [21,22] to oscillate with increasing Q up to very large Q
(see Fig. 22 of [21] and Fig. 2 of [22]). Note that while the description of
the multiphonon part of $S(Q,\omega)$ in terms of a peak position and half heights,
as in Fig. 3, is fairly reasonable at large Q, it is not really adequate in
the phonon-roton region when this component exhibits considerable structure
and long high frequency tails (Fig. 2).

The one-phonon intensity, $Z(Q)$, is nearly all the observed intensity at
small Q (< 0.4 Å^{-1}) where it increases approximately linearly with Q. It then
flattens off and passes through a shallow minimum in the maxon region before
rising to a strong maximum in the roton region after which it falls rapidly to
zero at $Q \approx 3.6$ Å^{-1}. The intensity of the multiphonon component increases
steadily with increasing Q until it saturates at the full observed intensity
beyond $Q \approx 3.6$ Å^{-1}. (For details see Fig. 12 of [21].)

We postpone discussion of some additional details of the pressure dependence of $S(Q,\omega)$ at low temperature until after we have discussed the temperature dependence of $S(Q,\omega)$ at SVP.

4. Temperature Dependence of $S(Q,\omega)$ at SVP

The earliest studies of the temperature dependence of the neutron inelastic
scattering by liquid ^{4}He at SVP were carried out by YARNELL et al. [3],
LARSSON and OTNES [6] and HENSHAW and WOODS [23,24]. Subsequent studies in
which there has been at least some systematic investigation of the temperature
dependence at SVP or other lowish pressure (e.g., 1 atm) have been carried out
by WOODS [25], COWLEY and WOODS [21], DIETRICH et al. [26], SVENSSON et al.
[17], WOODS and SVENSSON [15], BLAGOVESHCHENSKII and DOKUKIN [27], TARVIN and
PASSELL [28], MEZEI [29], MEZEI and STIRLING [30] and GOLUB et al. [31].

These studies, most of which have focused primarily on rotons, have shown
that the frequencies, $\omega(Q,T)$, and intrinsic half-widths, $\Gamma(Q,T)$, of the one-
phonon excitations decrease and increase, respectively, as the temperature is
raised in the superfluid phase, with the changes becoming increasingly more
rapid as one approaches T_λ. The early studies of HENSHAW and WOODS [23,24]
showed clearly that there was a marked change at T_λ in the slope of the
curves of width and mean energy (taken from the whole observed distributions),
with a relatively much weaker dependence on T above T_λ. The results of
DIETRICH et al. [26] showed similar behavior at all higher pressures. However, none of the studies carried out prior to 1978 gave evidence for any
qualitative change in the character of $S(Q,\omega)$ on passing through T_λ. This
was puzzling. One felt that the presence of the Bose condensate in the superfluid phase should have a significant effect on the dynamics and hence that
there should be a qualitative difference in $S(Q,\omega)$ above and below T_λ.
Further puzzles were presented by the facts that, except at rather low temperatures, the values of $\omega(Q,T)$ for rotons obtained from the neutron measurements
were in serious disagreement with the values inferred from thermodynamic
measurements, and the corresponding values of $\Gamma(Q,T)$ were in serious disagreement with the widths calculated using the theory of LANDAU and KHALATNIKOV
[32] if the parameters in the calculations were fixed using experimental thermodynamic data. The early calculations of thermodynamic quantities by BENDT
et al. [33], based on the neutron measurements of YARNELL et al. [3] for
temperatures up to 1.8 K, gave reasonably good agreement with the known
experimental values, but the later calculations of DIETRICH et al. [26] showed
that serious discrepancies appeared at temperatures within about 0.4 K of

T_λ. The early neutron studies [6,23,24,26] appeared to indicate a quite
dramatic decrease, 40-60%, in the roton gap, Δ, between low temperature and
T_λ at SVP or 1 atm pressure. In their extensive calculations of thermodynamic properties, BROOKS and DONNELLY [34] found that the effective thermal
roton gap, Δ_t, needed to give agreement with thermodynamic values was
systematically higher (e.g., by $\approx$ 25% at 2.1 K) than the values of Δ inferred
from the neutron experiments.

LARSSON and OTNES [6] found that their roton widths were some 3-4 times
larger than the values given by the Landau-Khalatnikov theory. Later, HENSHAW
and WOODS [23,24] and DIETRICH et al. [26] showed that reasonable agreement
with the widths from the neutron experiments could be obtained if one used in
the Landau-Khalatnikov calculations the temperature dependent values of Δ
inferred from the neutron experiments rather than parameters determined from
thermodynamic data. This procedure seemed to be a sort of self-consistent
treatment of the neutron results, but, unlike the comparison made by Larsson
and Otnes, it was not really an honest comparison with the Landau-Khalatnikov
theory and it was not very satisfying since the parameters used in the calculations were totally inconsistent with what was known from thermodynamic
measurements.

On the one hand, it was not really surprising to find some disagreements of
the type noted above. As the temperature is raised toward T_λ, the population of thermal rotons increases causing the roton energies and widths to
decrease and increase, respectively, thus leading to more and more poorly
defined excitations. If the roton widths and energies inferred from the early
neutron measurements [6,23,24,26] really were the correct values, then, near
T_λ, the full width, 2Γ, was approximately twice the excitation energy, Δ,
indicating very poorly defined excitations. Under these conditions, the
Landau-Khalatnikov theory and the procedures used to obtain Δ values from the
thermodynamic measurements would certainly not be expected to be strictly
valid. On the other hand, serious disagreements with the Landau-Khalatnikov
widths and the thermodynamic Δ values appeared at much lower temperatures than
one might have expected, i.e. at temperatures where the rotons still were
rather well defined excitations, and the disagreements, particularly the
factor of 3-4 disagreement with the Landau-Khalatnikov widths, were surprisingly large. One was thus inclined to suspect that the $\Gamma(Q,T)$ and $\omega(Q,T)$
values inferred from the neutron measurements were not the true single-
excitation widths and energies. These inferred values had been obtained by
analysing the whole observed distributions as if they corresponded to only
single-excitation scattering. At most, allowance was made for the instrumental resolution and for creation and annihilation processes. No allowance was
made for the multiphonon component of the scattering (see Fig. 2). To be
fair, the existence of this component was not known during the earliest
studies [6,23,24], but by the mid-seventies it had been extensively studied
[20,21,35-37] and it was clear that its presence needed to be allowed for in
the analysis. In addition, the distributions observed in the early studies
were, because of the way in which the measurements had been carried out, not
good representations of $S(Q,\omega)$ as evidenced by the fact that the multiphonon
scattering was totally missed. Both better experiments and better analysis
procedures were clearly needed.

While one could thus at least roughly understand the disagreements with the
Landau-Khalatnikov widths and the thermodynamic Δ values, there was still the
puzzle of why there seemed to be no significant change in the character of
$S(Q,\omega)$ on passing through T_λ. Was something being missed, or was the Bose
condensate really playing at most a very minor role in determining the
dynamics, and did rotons, which had originally been thought to be associated

with the superfluid phase, really persist, admittedly as rather ill-defined
excitations, to temperatures well above T_λ?

High resolution measurements reported in 1978 by SVENSSON, SCHERM and WOODS
[17] and WOODS and SVENSSON [15] finally showed that there was indeed a qualitative change in $S(Q,\omega)$ on passing through T_λ. Their results for the resolution broadened $S(Q,\omega)$ at the maxon position, $Q = 1.13$ Å^{-1}, are shown in
Fig. 4. We see that the sharp one-phonon peak which dominates the scattering
at T = 1.00 K (note the logarithmic scale) steadily broadens and becomes
weaker as the temperature is raised in the superfluid phase, but remains a
clearly identifiable feature in $S(Q,\omega)$ even at T = 2.15 K (just 0.02 K below
T_λ). Above T_λ there is, however, no evidence of this sharp feature; one
then has only the single broad peak characteristic of the scattering by non-
superfluid ^{4}He. Note also that there is little further change with increasing
temperature except at the low frequency edge where detailed-balance effects
are important and where there may also be a significant contribution, at least
at small wave vectors, from the central Rayleigh component [38,39] which gets
increasingly stronger as the temperature is raised above T_λ.

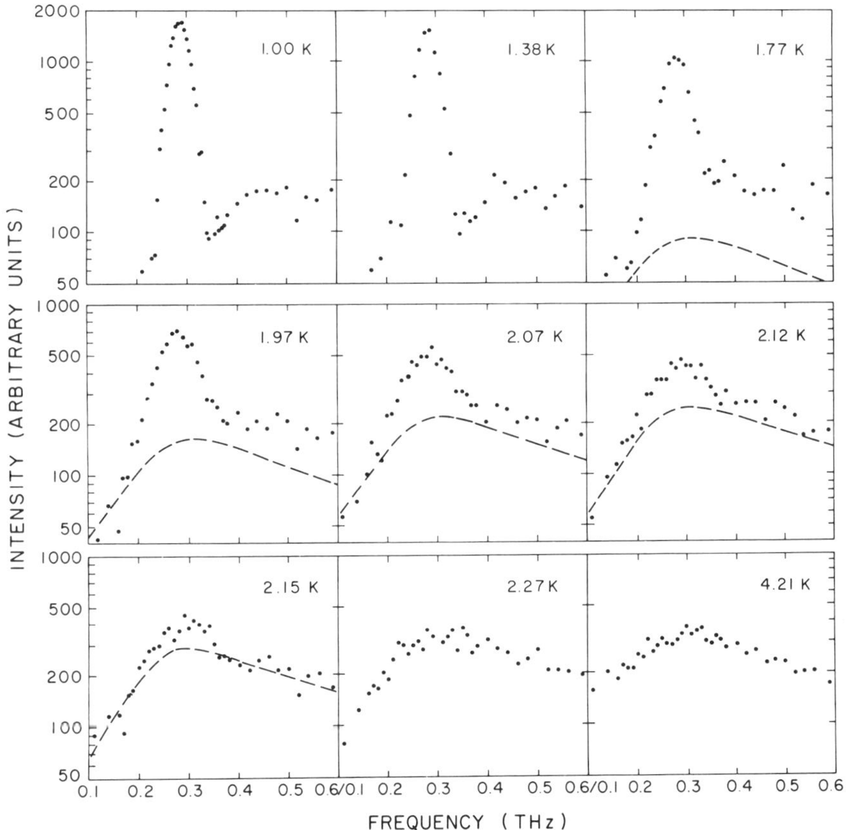

Fig. 4. Resolution-broadened dynamic structure factors for liquid ^{4}He for
$Q = 1.13$ Å^{-1} at SVP. The dashed curves show the "normal-fluid" components,
$(\rho_N/\rho)S_N(Q,\omega)$. From WOODS and SVENSSON [15]

The sharp peak in $S(Q,\omega)$ is thus unique to the superfluid phase and hence a signature of the presence of the Bose condensate. That this important qualitative difference in $S(Q,\omega)$ above and below T_λ was missed in the earlier studies [6,23,24,26] as well as some of the later ones [27,28] can readily be attributed to the ways in which the measurements were carried out - i.e., either with time-of-flight techniques or with fixed incident-neutron energy on a triple-axis spectrometer. In both cases, the high frequency part of $S(Q,\omega)$ is so strongly suppressed that the change in slope where the sharp peak intersects the broad component (in the region between 0.34 and 0.38 THz in Fig. 4) would not be detected, and hence the "disappearance" of the sharp peak at or very near to T_λ would not be noticed. It is clearly very important to carry out measurements in such a way as to obtain results that are a close approximation to the true resolution broadened $S(Q,\omega)$, as are those in Fig. 4.

The fact that the sharp "one-phonon" peak in $S(Q,\omega)$ seemed to disappear at T_λ led WOODS and SVENSSON [15] to propose a two-component model for $S(Q,\omega)$. In this model, which we shall henceforth refer to as the WS model, $S(Q,\omega)$ is envisaged as consisting of a "superfluid" component with a weight ρ_S/ρ and a "normal-fluid" component with a weight ρ_N/ρ, namely

$$S(Q,\omega) = \frac{\rho_S}{\rho}\, S_S(Q,\omega) + \frac{\rho_N}{\rho}\, S_N(Q,\omega). \tag{1}$$

Here ρ_S and $\rho_N = \rho - \rho_S$ are the temperature dependent macroscopic superfluid and normal-fluid densities. This model is, in essence, a two-fluid model for the scattering. The quantity $S_N(Q,\omega)$ represents the scattering characteristic of the normal fluid. In practice, its basic shape is determined by drawing a smooth curve through the experimental results for a temperature above T_λ (where $\rho_S = 0$ and $\rho_N = \rho$), but close to T_λ so as to avoid any complications by Rayleigh scattering. Woods and Svensson obtained $S_N(Q,\omega)$ from their results at 2.27 K but, in calculating the quantity $(\rho_N/\rho)S_N(Q,\omega)$ for $T < T_\lambda$, they included a thermal-population factor (i.e., a Bose function) to slightly adjust the shape at small ω as needed to satisfy detailed balance. The dashed curves in Fig. 4 show the values of $(\rho_N/\rho)S_N(Q,\omega)$ thus obtained.

At low temperature, where $\rho_N \approx 0$ and $\rho_S \approx \rho$, one observes directly the quantity $S_S(Q,\omega)$ which represents the scattering characteristic of the superfluid. Since $\rho_S = 0.993\,\rho$ at $T = 1.00$ K and SVP [34], the results for $T = 1.00$ K in Fig. 4 show essentially $S_S(Q,\omega)$ for 1.00 K, broadened of course by the experimental resolution. At this temperature, the intrinsic half-width of the one-phonon peak is [29] < 0.01 K ($< 2 \times 10^{-4}$ THz) so the intrinsic $S_S(Q,\omega)$ can be thought of as a δ-function one-phonon peak with a weight $Z(Q)$, i.e. $Z(Q)\delta[\omega-\omega(Q)]$, plus a broad multiphonon component, $S_M(Q,\omega)$, as originally proposed by MILLER et al. [40] for $T = 0$. At higher temperatures, the one-phonon peak begins to broaden appreciably and one must assume a form for the intrinsic line shape. Woods and Svensson found that a Lorentzian line shape gave a good description of their results. One can then write

$$S_S(Q,\omega) = \frac{Z(Q,T)}{\pi} \cdot \frac{\Gamma(Q,T)}{\Gamma^2(Q,T) + [\omega - \omega(Q,T)]^2} + S_M(Q,\omega). \tag{2}$$

Ignoring the energy-gain one-phonon peak at $\omega = -\omega(Q,T)$ is a rather good approximation at SVP where $\hbar\omega(Q,T) \gg k_B T$ even for the roton modes. In the WS model, essentially all of the negative-energy scattering observed at SVP is included in $S_N(Q,\omega)$.

The WS model, (1), can be applied in two ways: (a) One can subtract the
quantity $(\rho_N/\rho)S_N(Q,\omega)$ from the observed $S(Q,\omega)$ and then analyse the
remainder to obtain the "one-phonon" intensities, intrinsic widths and
frequencies via (2) or the equivalent expression for some other assumed form
for the one-phonon lineshape. (b) One can calculate $(\rho_S/\rho)S_S(Q,\omega)$ from
(2), using the observed scattering at the lowest temperature (e.g., T =
1.00 K) and some assumptions regarding the temperature dependences of the
various quantities, and combine it with $(\rho_N/\rho)S_N(Q,\omega)$ to obtain $S^{WS}(Q,\omega)$
for comparison with the observed $S(Q,\omega)$ at higher temperatures in the superfluid phase. In both cases one must of course make proper allowance for the
experimental resolution which, for constant-Q measurements on a triple-axis
crystal spectrometer, may, to a very good approximation, be taken to have a
Gaussian lineshape with a frequency dependent width that can be both calculated from the parameters of the instrument and checked against the observed
widths of the one-phonon peaks at very low temperature (where there is negligible intrinsic width) and the frequency width of the elastic scattering by
vanadium (an incoherent scatterer). Let's first consider approach (a).

When the normal-fluid components, $(\rho_N/\rho)S_N(Q,\omega)$, shown by the dashed
curves in Fig. 4 are subtracted from the full observed distributions, one is
left with the "superfluid" components shown in Fig. 5. At each temperature we
see that there is a well defined peak plus a weak broad component of scattered
intensity extending to high frequencies. (Only a limited region of the range,
-0.04 to 1.65 THz, covered by the measurements [15,17] is shown in Figs. 4 and

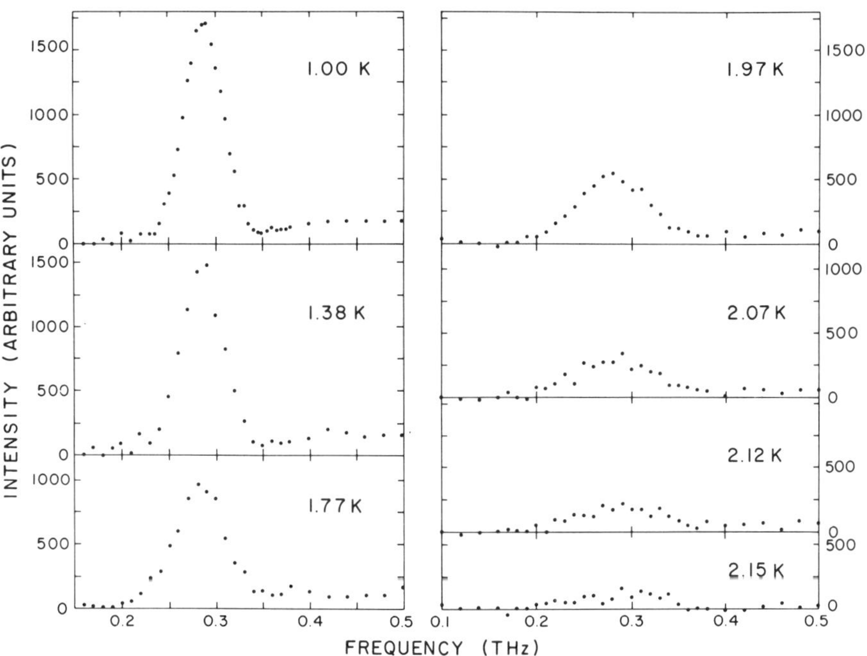

Fig. 5. The "superfluid" components for $Q = 1.13$ Å^{-1} obtained by subtracting
the "normal-fluid" components from the full experimental distributions shown
in Fig. 4. From WOODS and SVENSSON [15]

5.) Note that the intensity falls to zero on the low frequency side of the
peak and that the intensity in the broad component scales approximately as
that in the peak. Since, except for a temperature dependent broadening of the
main peak, all of the spectra in Fig. 5 have the same basic shape, one is led
to conclude, by analogy with what is well established at 1.0 K, that at each
higher temperature the "superfluid" spectrum still consists of a "one-phonon"
peak, which is symmetric within the experimental uncertainty, and a broad
"multiphonon" component. With simply-interpretable evolutions with temperature of the lineshapes of the postulated "superfluid" and "normal-fluid" components one can thus, as we shall see in more detail below, readily describe
the apparently complicated temperature dependence of the lineshape of the full
$S(Q,\omega)$ shown by the results in Fig. 4.

When Woods and Svensson analysed the results in Fig. 5, and the corresponding results for the four other Q values they studied, in terms of (2) (making,
in order to separate the one-phonon and multiphonon components, the usual
assumption [20] that the one-phonon peak is symmetric), they obtained the
"one-phonon" widths and reduced intensities shown in Fig. 6. Note that the
results for all five Q values appear to fall on universal curves. The
intensities scale essentially as ρ_S/ρ and there is excellent agreement
between the experimental widths and the widths (dashed curve) predicted by the
Landau-Khalatnikov theory [32]. The Landau-Khalatnikov widths were calculated
from the relation $\Gamma_{LK}(T) = 1.14\ T\ \rho_N/\rho$ where Γ and T are both in K.
Experimental values [41] for ρ_N/ρ were used and the numerical coefficient

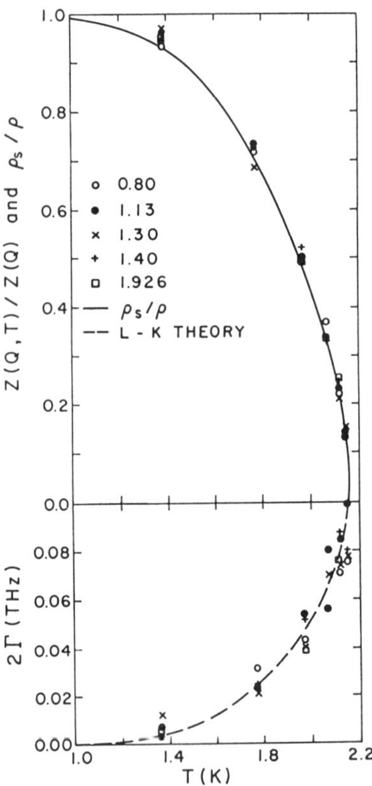

Fig. 6. Reduced intensities (normalized to ρ_S/ρ at 1.00 K) and
intrinsic full widths of one-phonon
excitations in superfluid ^{4}He at
SVP (for five Q values) obtained by
application (a) of the WS model
(see text). The solid curve shows
ρ_S/ρ and the dashed curve shows
the widths for rotons calculated
from the theory of LANDAU and
KHALATNIKOV [32]. From WOODS and
SVENSSON [15]

was determined using experimental viscosity values. There was no scaling of
$\Gamma_{LK}(T)$ to the experimental widths and no explicit account was taken of the
temperature variation of $\omega(Q,T)$ which, as we recall, was necessary to get even
approximate agreement with the values of $2\Gamma(Q,T)$ inferred from the earlier
neutron scattering studies [23,24,26]. The comparison of Fig. 6 was the first
"honest" comparison with the Landau-Khalatnikov theory to be made since the
work of LARSSON and OTNES [6] nearly 20 years earlier. As we noted earlier,
their widths, obtained from the full observed distributions, were 3-4 times
larger than the values given by the Landau-Khalatnikov theory. In contrast,
the intrinsic widths of the one-phonon peaks in the "superfluid" component
obtained by analysing $S(Q,\omega)$ in terms of the WS model, (1), are in essentially
perfect agreement with the Landau-Khalatnikov widths.

The agreement for the widths in Fig. 6 is in fact much better than one
would have anticipated. One would not have expected the Landau-Khalatnikov
theory, a perturbation type of theory, to be valid to such high temperatures,
where the excitations have quite substantial intrinsic widths. Note, however,
that, even just below T_λ, $2\Gamma(Q,T)$ is still only $\approx 0.5\ \omega(Q,T)$ for rotons and
$\approx 0.3\ \omega(Q,T)$ for maxons (see Fig. 17), so the "one-phonon" excitations, as
obtained from the WS model, are in fact reasonably well defined excitations
throughout the whole superfluid range. In contrast, the earlier studies [6,
23,24,26] indicated very poorly defined roton "excitations" near T_λ ($2\Gamma \approx$
2Δ, as mentioned previously) since the "roton widths" inferred from the whole
distributions in these studies were 2-3 times larger than the widths in Fig.
6, and the roton frequencies were also much lower (typically 30-50% lower near
T_λ) than the values obtained by WOODS and SVENSSON [15,42]. These lower
values partly reflect the fact that, for the roton wave vector, the "normal-
fluid" component (which is not removed when the whole distributions are used
in the analysis) is peaked at a frequency lower than that of the one-phonon
peak in the "superfluid" component (see Fig. 7). As mentioned previously, the
roton frequencies inferred from the early studies [6,23,24,26] exhibited a
decrease of 40-60% between low temperature and T_λ and, especially near T_λ,
were in serious disagreement with the values inferred from thermodynamic
measurements. In contrast, the analysis of WOODS and SVENSSON [15,42]
indicates that the roton frequency decreases by only about 13% between $T = 0$
and $T = T_\lambda$ (see Fig. 17) and gives values in agreement with those inferred
[34] from thermodynamic measurements.

Independent of any analysis procedure, one can see directly from Fig. 4
that the sharp "peak" in $S(Q,\omega)$, which undoubtedly corresponds to a one-phonon
excitation at the lowest temperature, persists as a separately identifiable
feature for all temperatures below T_λ. If, as seems reasonably likely to be
the case, this feature continues to correspond to the one-phonon excitation as
the temperature is raised, then these excitations clearly remain rather well
defined at least to within 0.02 K of T_λ. It would certainly not be reasonable to try and fit a function characterized by a single peak to any of the
full lineshapes for $T < T_\lambda$ shown in Fig. 4. There is clearly more than one
component for all $T < T_\lambda$ and a separation into the different components is
essential for a meaningful analysis. The WS model is of course only one of a
possibly large number of ways of achieving this separation, but we will
restrict our attention largely to this model in the present paper. This model
works rather well, at least as an empirical description of $S(Q,\omega)$ for all
$T < T_\lambda$, and, as we have already partly seen, it gives values for the
one-phonon parameters that resolve essentially all of the previously puzzling
disagreements between the results from neutron scattering measurements and
those from thermodynamic measurements.

Before proceeding further, we should note that, as has been pointed out by MINEEV [43], if the static structure factor, $S(Q) = \int S(Q,\omega)d\omega$, is independent of temperature, then the total weight, $S_S(Q) = \int S_S(Q,\omega)d\omega$, in the superfluid component obtained by subtracting $(\rho_N/\rho)S_N(Q,\omega)$ from the observed $S(Q,\omega)$ necessarily must scale as ρ_S/ρ. This fact, which was realized by Woods and Svensson, tells us nothing however about the shape of $S_S(Q,\omega)$ or the distribution of intensity between its "one-phonon" and "multiphonon" parts. It was the fact that the $S_S(Q,\omega)$ obtained by application (a) of the WS model had shapes (Fig. 5) that were very similar for all $T < T_\lambda$, and seemed to be readily understandable by a straightforward extension of what was known at low temperature, that led Woods and Svensson to believe that their model might well be a useful procedure for extracting the one-phonon-excitation part from the total $S(Q,\omega)$ at higher temperatures in the superfluid phase.

Let us now turn to application (b) of the WS model. As we noted earlier, the WS model is basically a two-fluid model for the neutron scattering by superfluid ^{4}He. It is clear from results such as those shown in Fig. 4 that the neutron scattering by the pure superfluid (essentially what is observed at $T = 1.00$ K) is qualitatively very different from the scattering by the pure normal fluid ($T > T_\lambda$). While two-fluid models are not normally used to describe microscopic properties, it does not seem totally unreasonable to expect that the scattering at intermediate temperatures might be some appropriately weighted combination of the scattering characteristic of the superfluid and that characteristic of the normal fluid, with due allowance of course for the evolution with temperature of the detailed lineshape of each. The scattering above T_λ (i.e., that characteristic of the normal fluid) is relatively insensitive to temperature although there are systematic changes in its leading (low frequency) edge, as required to satisfy detailed balance, and, especially as T aproaches 4.2 K and for low Q, changes brought about by the Rayleigh scattering [38,39] centred at $\omega = 0$. It seems plausible that the scattering characteristic of the normal fluid should persist below T_λ, with a weight given at least approximately by the amount of normal fluid present, i.e. ρ_N/ρ, and with little change in shape other than the sharpening up of the leading edge. Similarly, one might expect the scattering characteristic of the superfluid to persist to higher temperatures with a weight given approximately by the superfluid fraction, ρ_S/ρ, but with changes in the width and possibly also in the position of the one-phonon peak. By analogy with phonons, or perhaps even more appropriately magnons, in solids one would expect the intrinsic widths of the one-phonon excitations to increase as the temperature is raised and the frequencies to decrease. For rotons, we even have estimates (which should certainly be valid if the temperature is not too high) of the widths and frequency changes from the very old theory of LANDAU and KHALATNIKOV [32] and from the more recent work of BEDELL et al. [44,45].

In using their model, (1), to calculate $S(Q,\omega)$ for intermediate temperatures from the results observed at 1.00 and 2.27 K, WOODS and SVENSSON [15,42] calculated the normal-fluid component, $(\rho_N/\rho)S_N(Q,\omega)$, exactly as described earlier. In calculating the superfluid component, $(\rho_S/\rho)S_S(Q,\omega)$, they assumed an intrinsic Lorentzian lineshape, (2), for the one-phonon peak, with a width (assumed to be independent of Q as per Fig. 6) given by the Landau-Khalatnikov theory, and convoluted this Lorentzian with the known Gaussian experimental resolution. They assumed that the shape of the broad multiphonon component, $S_M(Q,\omega)$, was independent of temperature but that, like the one-phonon peak, its intensity scaled as ρ_S/ρ. The shape of $S_M(Q,\omega)$ may of course change as the single excitations broaden and shift in frequency with increasing temperature but the changes would be expected to be relatively much smaller than those of the one-phonon peak, and, at present, there is no theory

to guide one in estimating the changes. At any rate, the contribution of
$S_M(Q,\omega)$ to the total $S(Q,\omega)$ becomes increasingly less important as the
temperature increases (see Fig. 5) so minor variations in its lineshape are of
little consequence. Woods and Svensson also assumed that the one-phonon
frequency, $\omega(Q,T)$, was independent of temperature. It was not essential to do
so since they could have incorporated the Q-dependent frequency variation
inferred from application (a) of their model (see Fig. 17) or, for rotons,
calculated the temperature variation of Δ from the Landau-Khalatnikov theory.
However, if one allows $\omega(Q,T)$ to vary with T then one also needs to allow the
intensity to vary as required to keep the first moment constant, and this
extra level of complication did not seem to be merited for the initial testing
of a purely empirical model.

In addition to the explicit assumptions regarding the intensities and
lineshapes of the different components, the WS analysis also implicitly
assumes that the density, ρ, and the integrated intensity of $S(Q,\omega)$, i.e.
$S(Q)$, are independent of temperature for $T < T_\lambda$. The density does indeed
only vary by a small amount ($\approx 0.7\%$) for $T < T_\lambda$ at SVP. $S(Q)$ is known [46-
48] to be relatively but certainly not totally insensitive to temperature for
$T < T_\lambda$ at SVP. There is very little dependence on temperature for Q in the
maxon region (see Fig. 1 of [47]), but for Q in the roton region there is a
decrease of about 5% between T_λ and 1.0 K, with a rapid drop below T_λ
characteristic of an order-parameter type of temperature dependence (see Fig.
3 of [48]). One would thus expect the WS model to give a poorer description
of $S(Q,\omega)$ for rotons than for maxons, even more so when one notes that the
roton frequency is much more sensitive to temperature than the maxon frequency
(see Fig. 17).

In Fig. 7, we show, for $Q = 0.80$ Å^{-1} and $Q = 1.926$ Å^{-1} (the roton minimum
at SVP) at $T = 2.12$ K (where $\rho_S = 0.2\,\rho$), comparisons [42] of the experimental distributions and the predictions (solid curves) of the WS model. The
normal-fluid components, $(\rho_N/\rho)S_N(Q,\omega)$, are shown as dashed curves. At
$Q = 0.8$ Å^{-1}, where $S(Q)$ and $\omega(Q,T)$ are essentially independent of temperature,
the WS model gives an excellent description. Similarly good agreement is
obtained [42] for $Q = 0.8$ Å^{-1} at other temperatures and for $Q = 1.13$ (the
maxon), 1.3 and 1.4 Å^{-1}. At $Q = 1.926$ Å^{-1}, where $S(Q)$ increases by about 5%

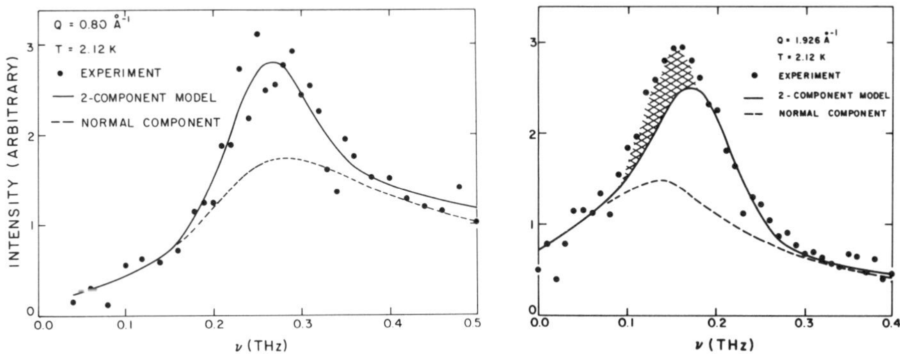

Fig. 7. Comparison of the experimental resolution-broadened dynamic structure
factors for $Q = 0.8$ Å^{-1} (left) and $Q = 1.926$ Å^{-1} (right) at $T = 2.12$ K and
SVP with the distributions (solid curves) obtained by application (b) of the
WS model (see text). The dashed curves show the normal-fluid components,
$(\rho_N/\rho)S_N(Q,\omega)$. From WOODS and SVENSSON [42]

between 1.0 K and T_λ and $\omega(Q,T)$ decreases by about 13%, the description is
still very good over most of the frequency range, but is inadequate in the
region of the sharp peak. The WS model, as applied here, allows the sharp
peak to broaden but not shift in frequency as the temperature increases so it
gets the position of the roton peak too high in frequency at 2.12 K. It also
gets the intensity too low. The intensity in the roton peak seems not to
decrease quite as rapidly with increasing temperature as the quantity ρ_S/ρ.
We know of course that the intensity change that gives rise to the substantial
increase in $S(Q)$ with increasing temperature must appear somewhere in $S(Q,\omega)$,
and the results shown in Fig. 7 suggest that much of it is concentrated in the
sharp one-phonon peak.

As we noted earlier, one could readily tinker with the WS model to get
better agreement with the experimental $S(Q,\omega)$ in application (b). Frequency
shifts and their accompanying intensity changes can readily be incorporated
and the weighting of the different components can be adjusted to make the
total intensity agree with the known $S(Q)$ at each temperature. This would
give much better agreement for the roton wave vector while making negligible
difference for the other wave vectors studied by Woods and Svensson, where the
agreement is already essentially perfect. Such tinkering seems hardly to be
merited, however, since, at the present time, the WS model is a purely empirical model. Its principal use is in extracting one-phonon widths and frequencies at higher temperatures. For this application, i.e. application (a), much
of the tinkering that one might do to get better agreement in application (b)
is largely irrelevant. All that matters in application (a) is what one
chooses for the normal-fluid component that one then subtracts from the full
experimental distribution to get the superfluid component.

It is interesting to make a detailed comparison of the roton widths and
frequencies obtained by WOODS and SVENSSON [15,42] with the values obtained in
other studies. MEZEI [29], using neutron spin-echo techniques, has determined
with very high accuracy the roton linewidths and frequency shifts at low
temperatures (< 1.4 K) where one would expect the Landau-Khalatnikov theory to
be valid. His results are shown in Fig. 8 together with the values obtained
by DIETRICH et al. [26], TARVIN and PASSELL [28] and WOODS and SVENSSON
[15,42] from neutron scattering experiments and the values obtained by GREYTAK
and YAN [49] from Raman scattering experiments. The solid curves show Mezei's
calculations based on the Landau-Khalatnikov theory and the dashed curve shows
the upper-bound on the roton frequency shift, $\Delta(0) - \Delta(T) < T$, given by the
theory of BEDELL et al. [44]. Mezei obtained the solid curves by first fitting the functional form given by the Landau-Khalatnikov theory to his results
(open circles) for $\Delta(0) - \Delta(T)$, taking into account the known value [50] of
$\Delta(0)$, and then using the values of $\Delta(T)$ thus obtained and the value, 47, of
the coefficient calculated from viscosity data to calculate the curve for the
widths. This is essentially the same procedure that was used by HENSHAW and
WOODS [23,24] and DIETRICH et al. [26] to obtain approximate consistency
between the widths inferred from their neutron experiments and those calculated using the Landau-Khalatnikov theory, and that we have criticized above as
not being an honest comparison with the Landau-Khalatnikov theory. Mezei's
calculations are, however, based only on results at low temperatures where
there is virtually no doubt (other than the uncertainty in the experimental
measurements) about the excitation widths and frequencies and no inconsistency
with what is known from thermodynamic measurements. The comparison shown in
Fig. 8 is thus an honest comparison with the Landau-Khalatnikov theory. Note,
however, that there is a possibility of systematic errors in the Landau-
Khalatnikov curves at high temperatures since the values of $\Delta(T)$ are determined by fitting to a very few experimental values, with rather large percent-
age errors, at low temperatures.

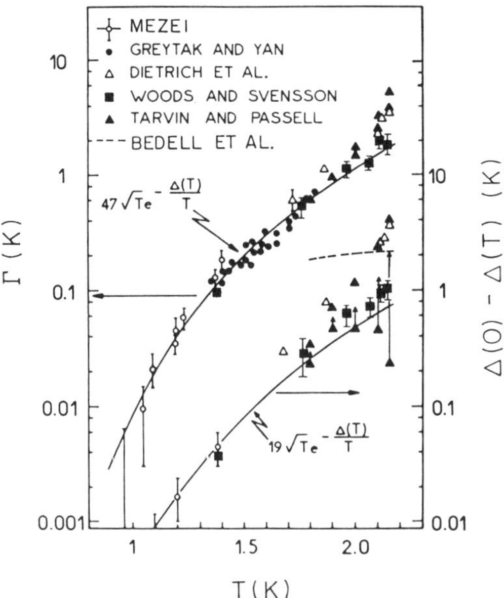

Fig. 8. Intrinsic widths and frequency shifts for rotons in superfluid ^{4}He at SVP obtained by MEZEI [29] (open circles and solid curves), GREYTAK and YAN [49] (solid circles). DIETRICH et al. [26] (open triangles), WOODS and SVENSSON [15,42] (solid squares) and TARVIN and PASSELL [28] (solid triangles). The dashed curve shows the bound on the shift from the theory of BEDELL et al. [44]. The tips of the arrows extending upward from the lower solid triangles for the shifts show the "corrected" Tarvin and Passell harmonic oscillator values (see text). Adapted from MEZEI [29]

As expected, the Landau-Khalatnikov theory gives an excellent description of Mezei's low temperature results. The Landau-Khalatnikov widths are also in very good agreement with the Raman results of Greytak and Yan and with the results of Woods and Svensson for temperatures right up to T_λ. There are, however, discrepancies of a factor of 2-3 (note the logarithmic scale) near T_λ with the results of Dietrich et al. and Tarvin and Passell. These discrepancies are attributable to a number of factors as we shall discuss below. The Landau-Khalatnikov curve for the frequency shift lies systematically below the values of Woods and Svensson at high temperatures. Considering that this curve was determined from a few results at temperatures below 1.4 K, the differences are, however, probably well within the combined uncertainties. One could readily construct a new Landau-Khalatnikov curve that would give good agreement with the WS shifts while still being consistent with Mezei's values at low temperatures. The shifts (triangles) obtained by Dietrich et al. and Tarvin and Passell differ from those of Woods and Svensson by as much as a factor of four (in both directions) near T_λ. Note also that, near T_λ, their shifts exceed, by as much as a factor of two, the bound on the shift (dashed curve) that is a consequence of a stability condition in the "roton liquid theory" of BEDELL et al. [44].

Note that TARVIN and PASSELL [28] give two values of the shift and width at each temperature. The smaller shifts and larger widths were obtained by analysing their spectra in terms of a harmonic oscillator model for the lineshape while the larger shifts and smaller widths were obtained by an analysis in terms of the same "Lorentzian" model as was used by DIETRICH et al. [26]. As has recently been noted by TALBOT et al. [19], the "Lorentzian" model used by Dietrich et al. and Tarvin and Passell is actually a Lorentzian multiplied by ω which is an inappropriate model for the one-phonon lineshape. The presence of the ω causes the values of $\omega(Q,T)$ and $\Gamma(Q,T)$ obtained by using this model to be distorted, with the distortion becoming increasingly more serious as the width increases. Results obtained by use of this model should thus be rejected. Talbot et al. have also pointed out that Tarvin and Passell made an incorrect identification of the excitation energy for their harmonic oscillator model. Instead of $\omega(Q,T)$ they used the quantity $E(Q,T)$, given by $E^2(Q,T) = \omega^2(Q,T) + \Gamma^2(Q,T)$, as the excitation energy. Their harmonic oscillator "excitation energies" are thus too large near T_λ, and hence the corresponding energy shifts (the values shown by the lower solid triangles in Fig. 8) are much too small. One can readily "correct" the Tarvin and Passell harmonic oscillator results by using their tabulated values of $E(Q,T)$ and $\Gamma(Q,T)$ to obtain values of $\omega(Q,T)$. Using these $\omega(Q,T)$ values and normalizing to Mezei's shift curve at 1.34 K, the lowest temperature of the Tarvin and Passell measurements, gives "corrected" harmonic oscillator shifts which are indicated by the tips of the arrows extending upward from the lower set of solid triangles in Fig. 8 (see also Fig. 12 of [19]). These "corrected" shifts are in very good agreement with the WS values for temperatures up to 2.0 K but then deviate upward and, at 2.15 K, are a factor of 2 larger than the WS value. The deviation near T_λ for both the shifts and the widths (see Fig. 8) is attributable to the fact that Tarvin and Passell analysed their full observed distributions as if they corresponded to single-excitation scattering while Woods and Svensson fitted only to what they identified as the "one-phonon" peak in the "superfluid" component of $S(Q,\omega)$. While one may argue about precisely how to separate $S(Q,\omega)$ into different components, it is obvious from the experimental results (Fig. 4) that $S(Q,\omega)$ for superfluid ^{4}He consists of more than one component and a separation into different components is essential for a meaningful analysis. Treating the full observed distributions as if they corresponded to single-excitation scattering, as was done by TARVIN and PASSELL [28], DIETRICH et al. [26] and earlier authors [6,23,24], cannot, even with proper allowance for both energy-gain and energy-loss processes, give reliable results for $\omega(Q,T)$ and $\Gamma(Q,T)$ as T approaches T_λ. One should note that the harmonic oscillator model used by Tarvin and Passell is simply the combination of an energy-loss Lorentzian centred at $\omega = \omega(Q,T)$, i.e. the Stokes term, and the corresponding energy-gain Lorentzian centred at $\omega = -\omega(Q,T)$, i.e. the anti-Stokes term. (See the Appendix of [19] for a detailed discussion.) Since, as we noted in connection with (2), the energy gain one-phonon peak can essentially be ignored at SVP, the harmonic oscillator model and the single-Lorentzian model, (2), used by Woods and Svensson would give the same answer if applied to the one-phonon component identified in the WS analysis. The differences between the WS parameters and the "corrected" harmonic oscillator parameters of Tarvin and Passell are thus due not to the different models employed but rather to the different data to which the models were applied - to the whole distribution in one case and to only that part identified as the sharp one-phonon component in the other.

As a rough measure of how much the $\omega \times$ Lorentzian model used by Dietrich et al. and Tarvin and Passell distorts the results, we call attention to the fact that the energy shifts that Tarvin and Passell obtained using this model (upper solid triangles in Fig. 8) are larger than the "corrected" harmonic oscillator shifts by factors that increase from 1.3 at 1.8 K to 1.75 at 2.0 K and 2.0 at 2.15 K. Since they were both applied to the full observed distributions, neither model of course gives correct one-phonon parameters for

temperatures near T_λ. Although the $\omega(Q,T)$ and $\Gamma(Q,T)$ values given by
Dietrich et al. and Tarvin and Passell are not reliable, especially at the
higher temperatures, there is no reason to question the validity of their
basic experimental results. It would in fact be very valuable to have these
results reanalysed, especially those of DIETRICH et al. [26] which are still
the only extensive set of results for the temperature dependence of rotons
over a wide range of pressures.

The predictions (solid curves) of BEDELL et al. [44,45] for the roton
widths and energy shifts are compared in Fig. 9 with the values of MEZEI [29]
and WOODS and SVENSSON [15,42] and with a much less complete selection of the
values of the other authors than was shown in Fig. 8. Note in particular that
the values of Dietrich et al. and Tarvin and Passell for the temperatures
nearest to T_λ have been omitted. There is seen to be generally good
agreement between the predictions of Bedell et al. and the results inferred
from the experiments. As was the case for the Landau-Khalatnikov widths in
Figs. 6 and 8, there is no significant difference between the widths predicted
by Bedell et al. and those from the WS analysis, right up to T_λ. The shifts
predicted by Bedell et al. lie systematically above the shifts obtained by
Mezei and Woods and Svensson but the differences are probably not outside the
combined uncertainties. In contrast, the Landau-Khalatnikov shifts in Fig. 8
were systematically lower than the WS values near T_λ, by approximately the
same amount as the predicted shifts in Fig. 9 are higher than the WS values.
Note again that the WS energy shifts fall well within the bound on the shift
(dashed curve) that results from a stability condition in "roton liquid
theory" [44].

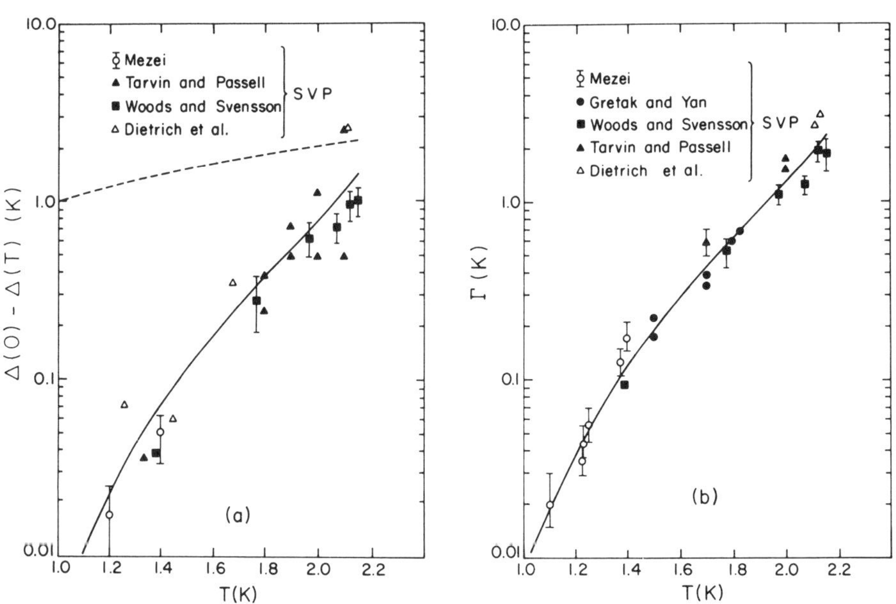

Fig. 9. Frequency shifts and intrinsic widths for rotons in superfluid [4]He at
SVP calculated by BEDELL et al. [45] (solid curves) and a selection of the
experimental values from Fig. 8. (Symbols and references are the same as in
Fig. 8.) The dashed curve shows the bound on the shift from "roton liquid
theory" [44]. Adapted from BEDELL et al. [45]

75

In the theories of Landau and Khalatnikov and Bedell et al., the excitations decay via 4-phonon processes and the temperature variations of Δ and Γ arise predominantly from the changing population of the thermal rotons needed for this decay process to occur. Since these theories are basically perturbation theories using infinite lifetime intermediate state phonons, one would not expect them to be strictly valid when/if the excitation lifetimes become short, i.e. $\Gamma(Q,T)$ becomes large. The good agreement with the experimental widths and frequency shifts, as determined via the WS analysis procedure, is thus somewhat unexpected and surprising. As we have noted previously, however, the sharp peak in $S(Q,\omega)$ which we have interpreted as corresponding to the one-phonon excitation, persists as a well defined entity (see Fig. 4) at least to 2.15 K (i.e., to within 0.02 K of T_λ). If this identification is correct, then, independent of any analysis procedure, the one-phonon excitations are rather well defined excitations even at temperatures very close to T_λ. The WS analysis procedure in fact indicates that, even for the roton, the full width at half maximum, $2\Gamma(Q,T)$, is only 0.5 $\omega(Q,T)$ at T = 2.15 K (see Figs. 17 and 18). With such well defined excitations one should perhaps not be so surprised to find the good agreement shown in Figs. 6 (lower half), 8 and 9. I take this good agreement as an indication that the WS analysis procedure has done a rather good job of identifying the one-phonon component of $S(Q,\omega)$ and that the theories of Landau and Khalatnikov and Bedell et al. are at least approximately valid throughout the whole superfluid phase.

The WS model, (1), has given us a rather simple interpretation of the apparently complicated temperature dependence of the total $S(Q,\omega)$ for superfluid ^{4}He and it has allowed us to extract values of $\omega(Q,T)$ and $\Gamma(Q,T)$ for the one-phonon excitations that, for all $T < T_\lambda$, are in very good agreement with theoretical predictions that make use of the results of thermodynamic measurements. The previously puzzling disagreements between the results from thermodynamic measurements and those from neutron scattering studies have been entirely eliminated. While this is a rather pleasing state of affairs, we emphasize that the WS model is a purely empirical model which had its origins in the observation [15,17] that there was a qualitative change in the character of $S(Q,\omega)$ on passing through the λ point, with a sharp peak being present at all temperatures below T_λ, but not above (see Fig. 4). This sharp peak, which appears to have a weight at least approximately proportional to the superfluid density, ρ_S/ρ, must be a signature of the presence of the Bose condensate.

There have been several attempts [43,51-54] to provide a theoretical basis for the WS model but, as has been emphasized by GRIFFIN [55], there is at present no theoretical justification for the model. The early theoretical studies by GRIFFIN [51] and GRIFFIN and TALBOT [52] appeared to give direct support for (1), but in more recent work [53] these authors find that, at all temperatures, $S(Q,\omega)$ exhibits only a single peak, centred at the one-phonon frequency and with a width which scales roughly as ρ_N/ρ. In contrast to what appears to be indicated by both the results of Woods and Svensson at SVP (Fig. 4) and the more recent results [18,19] at high pressure (see Figs. 13 and 14), their calculations give no indication of a sharp peak appearing on top of a broader one as one goes below T_λ. Note, however, that their calculations were for Q = 0.35 and 0.8 Å^{-1} and that the approximations they used were such that their calculations should only be strictly valid at sufficiently small Q that the multiphonon scattering is negligibly small.

In the empirical WS model, the phonon-roton modes have a weight proportional to ρ_S/ρ and hence do not exist above T_λ. This is not the case in the finite-temperature microscopic theory [53,55,56] of Griffin and co-workers. In their theory, the phonon-roton modes are envisaged as being

zero-sound density fluctuations existing both above and below T_λ. There is,
however, a marked difference above and below T_λ since, below T_λ, where
there is a coupling of density and field fluctuations because of the Bose
broken symmetry, the density fluctuation modes become poles in the single-
particle Green's function (i.e., elementary excitations of the system). One
thus anticipates a possible shift in the excitation energy and a change in the
decay mechanism (which determines the excitation width or inverse lifetime) on
crossing T_λ. This theory, which has clearly shown that the existence of the
Bose condensate is of crucial importance for the understanding of superfluid
^{4}He, can thus account, at least qualitatively, for a markedly different
behavior above and below T_λ, as observed in the neutron scattering
experiments [15,17-19]. (For additional discussion see [19,53,55,56].)

The study of WOODS and SVENSSON [15] gave the somewhat surprising result
(Fig. 6) that the excitation width was, to within the accuracy of the experiments, independent of Q over the range $0.8 \le Q \le 1.926$ Å^{-1}. This does not
continue to be the case at lower Q. The intrinsic linewidths for $0.3 < Q <$
0.7 Å^{-1} and $T < 1.7$ K obtained by MEZEI and STIRLING [30] under conditions
of very high experimental resolution are shown in Fig. 10. There is a strong
dependence on Q in this low Q region as was also evident in the earlier measurements of COWLEY and WOODS [21] (see their Fig. 17). The calculations of
TALBOT and GRIFFIN [53] also indicate a marked difference in linewidth for
$Q = 0.35$ Å^{-1} and $Q = 0.8$ Å^{-1}. Their calculated widths are, however, much
smaller than the corresponding values shown in Fig. 10 and Fig. 6. Note in
Fig. 10 that there is a drop in width between 0.5 and 0.6 Å^{-1} at 0.95 and 1.2
K, and probably also at 1.4 and 1.5 K. This undoubtedly reflects the fact
that 3-phonon decay processes, which are dominant in the region of anomalous
dispersion at least at low temperatures (see MARIS [57] for details), are no
longer allowed when one passes into the region of normal dispersion above
$Q = 0.55$ Å^{-1} [58]. It would be very valuable to have measurements at low Q,
of the quality of those of Mezei and Stirling (Fig. 10), extending to temperatures above T_λ.

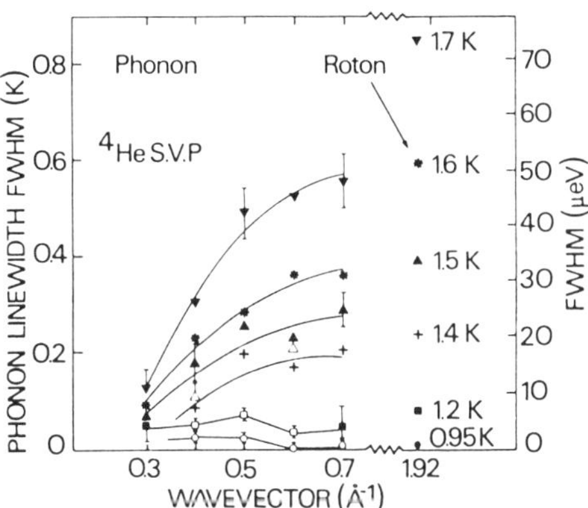

Fig. 10. Intrinsic full widths for rotons and low-Q phonons in liquid ^{4}He at
SVP for temperatures up to 1.7 K. From MEZEI and STIRLING [30]

We have already seen (Fig. 2) that the frequencies, widths and intensities of
the one-phonon peaks in $S(Q,\omega)$ at low temperature can be quite markedly
altered by applying pressure. Additional details of the effects of pressure
on $S(Q,\omega)$ at low temperature are summarized in Figs. 11 and 12 taken, respectively, from papers by SVENSSON and TENNANT [59] and GRAF et al. [37]. The
frequencies of most of the one-phonon excitations increase with increasing
pressure but those in the vicinity of the roton minimum and beyond decrease,
and the roton minimum also moves to larger Q. The dramatic decrease in the
intensity of the maxon peak between low and high pressure shown in Fig. 2 is
probably attributable to a greater interaction between the maxon, which has
moved up in frequency, and the multiphonon component which, having a large
two-roton contribution, has moved down in frequency. At pressures beyond
about 18 atm, the maxon frequency exceeds twice the roton frequency (Fig.

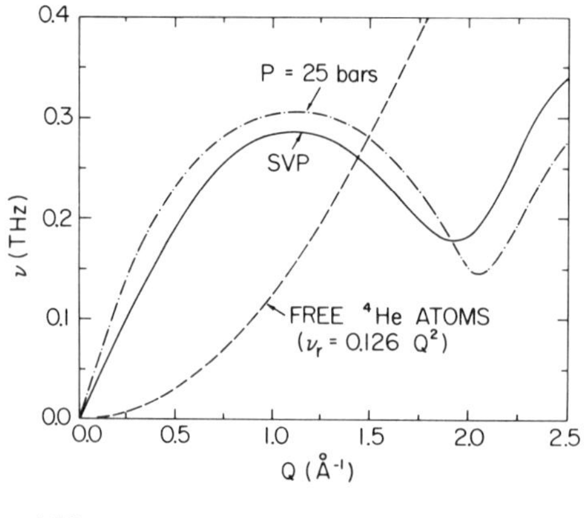

Fig. 11. The dispersion relations
for free ⁴He atoms and for one-
phonon excitations in low temperature superfluid ⁴He at SVP and 25
bars. From SVENSSON and TENNANT
[59]

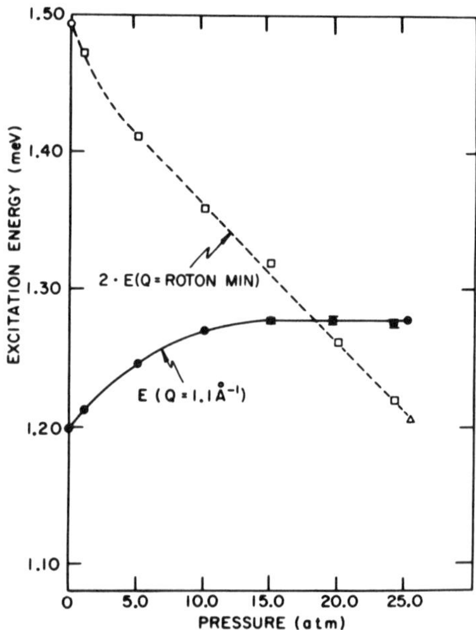

Fig. 12. The pressure dependence
of the maxon energy and twice the
roton energy for liquid ⁴He at low
temperature. Adapted from GRAF et
al. [37]

12). This opens another possible decay channel for the single excitation and may explain why the maxon peak at 24 bars in Fig. 2 has a significant intrinsic width, even at 1.2 K.

By applying pressure, one can thus make quite dramatic changes in the relative frequencies and intensities of the one-phonon peaks at different Q values and of the one-phonon and multiphonon components at a given Q value. Given these known effects, it was felt that detailed measurements of the temperature dependence of $S(Q,\omega)$ at a high pressure, similar to those [15,17] at SVP which we have just discussed, would be most valuable in providing additional critical tests of theories and models for $S(Q,\omega)$ and further insights on the differences in the normal and superfluid phases and hence on the role of the Bose condensate in determining the dynamics of superfluid ^{4}He. The results of such measurements at a pressure of 20 bars have recently been reported by SVENSSON et al. [18] and TALBOT et al. [19]. Measurements at 20 bars are in general somewhat more difficult than at SVP. At the maxon position it is less clear how to separate the one-phonon and multiphonon components even at the lowest temperature (see Fig. 2) and, since the one-phonon peak starts out much weaker at low temperature, it is more difficult to follow it to temperatures close to T_λ. Although the roton peak is slightly stronger than at SVP, it has moved to a substantially lower frequency making it more difficult to separate from the strong elastic scattering by the pressure cell (necessarily thicker than a cell designed for SVP) especially as the peak weakens, broadens and moves to even lower frequency as the temperature is raised toward T_λ. At 20 bars, the λ-point (1.928 K) is also considerably lower than at SVP (2.172 K) and ρ_S/ρ and ρ_N/ρ are quite different functions of temperature. Since ρ_S/ρ and ρ_N/ρ are, however, universal functions of T/T_λ (see Fig. 9 of [41]) this just implies a rescaling of the results, but the lower T_λ does make it more difficult to achieve small values of T/T_λ, and hence large values of ρ_S/ρ, since the lowest temperature is set by the experimental apparatus. In spite of these difficulties, the high flux available at the Institut Laue-Langevin, Grenoble allowed a rather complete set of high quality results to be otained for both the maxon $(Q = 1.13 \text{ Å}^{-1})$ and the roton $(Q = 2.03 \text{ Å}^{-1})$ under high resolution conditions very similar to those of WOODS and SVENSSON [15]. The measurements covered the range from 1.29 K to 3.94 K, with particular emphasis on temperatures below and just above T_λ.

Results for the resolution broadened $S(Q,\omega)$ at the maxon wave vector are summarized in Fig. 13. These curves were obtained by fitting the WS model to the observed $S(Q,\omega)$ (see [19] for details). This model gives an excellent description for all $T < T_\lambda$ for $Q = 1.13 \text{ Å}^{-1}$ (see Fig. 1 in [19]). As at SVP (Fig. 4), we see that the sharp one-phonon peak broadens and decreases rapidly in intensity as the temperature is raised, but changes very little in frequency up to the point where it becomes too weak to observe as a separate feature in $S(Q,\omega)$. We cannot follow this sharp feature as close to T_λ as at SVP but the indications are that, as at SVP, the sharp maxon peak at $P = 20$ bars just steadily decreases in intensity, while broadening only moderately and hardly changing in position, and disappears at T_λ. Above T_λ we just have the single broad peak characteristic of non-superfluid ^{4}He which, to within the experimental uncertainty, then remains unchanged to the highest temperature studied, 3.94 K (see Fig. 3 of [19]). Note that this normal-fluid peak is, for this Q value, centred at a frequency roughly 1.8 times higher than that of the sharp one-phonon peak observed below T_λ. Note also that there is no significant temperature dependence to the intensity above $\nu \approx 0.7$ THz. This is expected since numerous theoretical studies (see, e.g., [53], and also [13] for additional details and complete references) have shown that the high-frequency tail of $S(Q,\omega)$ should be independent of temperature and,

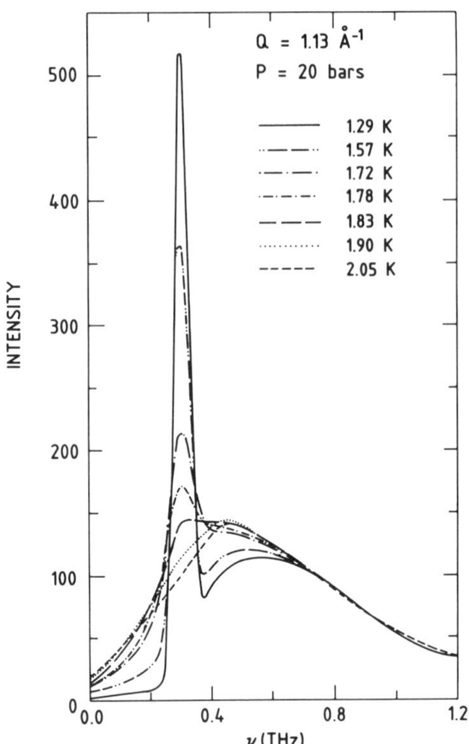

Fig. 13. Temperature dependence of the resolution-broadened dynamic structure factor of liquid ^{4}He for $Q = 1.13$ $\mathrm{\AA}^{-1}$ at $P = 20$ bars. These curves were obtained in the study of TALBOT et al. [19] by fitting the WS model to the experimental distributions

for a given Q, should exhibit an $\omega^{-7/2}$ frequency dependence. The precise frequency dependence of the tail has not been checked for the results of Fig. 13, but WONG [60] has shown that the results of WOODS et al. [35] for $Q = 0.8$ $\mathrm{\AA}^{-1}$ are consistent with the expected $\omega^{-7/2}$ dependence.

Results for the resolution broadened $S(Q,\omega)$ at the roton wave vector for the lowest temperature studied (1.29 K) and for six temperatures in the vicinity of the λ-point are shown in Fig. 14. The sharp and very strong one-phonon peak observed at 1.29 K decreases rapidly in intensity, broadens and shifts significantly to lower frequency as the temperature is raised. We can still clearly see this peak at 1.90 K but it then disappears as we go above T_λ (1.928 K) and we are again left with just the single broad peak (dashed curve in Fig. 14) characteristic of non-superfluid ^{4}He. Note that there is still a hint of the one-phonon peak (i.e., net intensity above the dashed curve in the region $\nu \approx 0.13$ THz) at $T = 1.93$ K. This temperature is nominally above T_λ, but the temperatures in this study [18,19] were only believed to be accurate to $\pm$ 0.02 K so the true temperature may still be slightly below T_λ. Since a constant-flow cryostat was used for the measurements, no in situ temperature calibration using the known λ-point could be made. (Such a calibration was made in the study of WOODS and SVENSSON [15] at SVP discussed in the previous section, and the temperatures quoted there are accurate to better than $\pm$ 0.005 K near T_λ and $\pm$ 0.01 K everywhere.) Note that, at the roton wave vector, the broad peak observed above T_λ is centred at a frequency about half that of the one-phonon peak at low temperature, in marked contrast to the situation

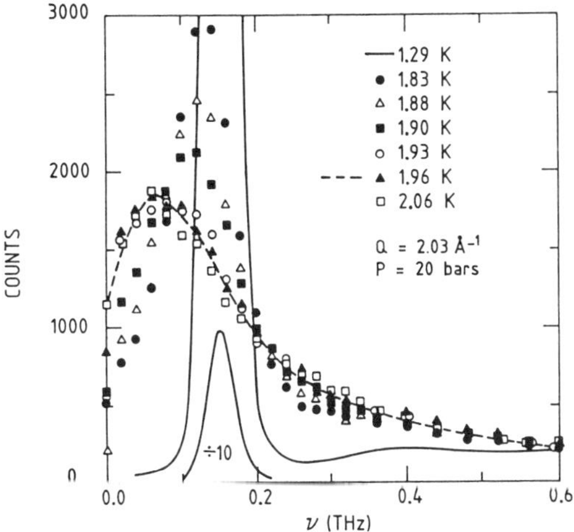

Fig. 14. Temperature dependence of the resolution-broadened dynamic structure factor of liquid ^{4}He for $Q = 2.03$ Å^{-1} at $P = 20$ bars. Adapted from SVENSSON et al. [18]

at the maxon wave vector (Fig. 13). The normal-fluid peak is also more sensitive to temperature change above T_λ for $Q = 2.03$ Å^{-1} than for $Q = 1.13$ Å^{-1} (see Fig. 3 of [19]) as would be expected (because of the requirements of detailed balance) for a peak centred at such low frequency that it has considerable intensity in the region $\nu < 0$.

Examples of the descriptions of the observed distributions in terms of the WS model are shown in Fig. 15 for each of the Q values studied. These results

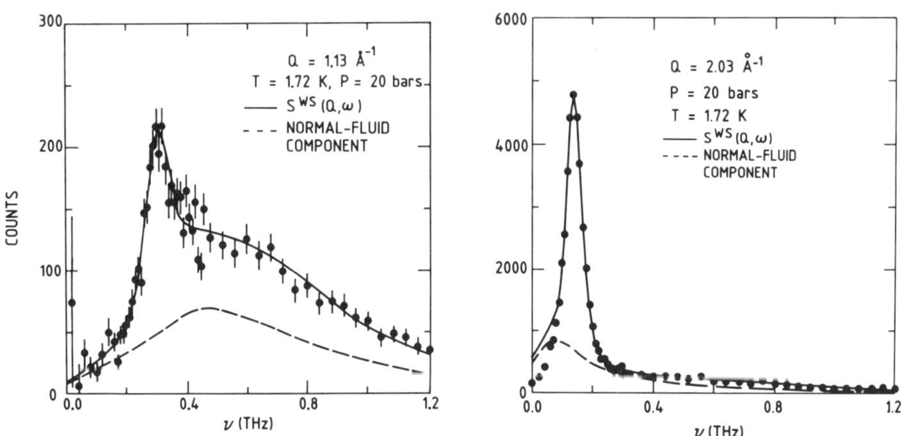

Fig. 15. Comparison of the experimental resolution-broadened dynamic structure factors for $Q = 1.13$ Å^{-1} (left) and $Q = 2.03$ Å^{-1} (right) at 1.72 K and $P = 20$ bars with the distributions (solid curves) obtained by fitting the WS model to the experimental spectra (see text). The dashed curves show the normal-fluid components, $(\rho_N/\rho)S_N(Q,\omega)$. Results taken from the study of TALBOT et al. [19]. Left figure adapted from SVENSSON et al. [18]

are for T = 1.72 K where ρ_S = 0.51 ρ. The dashed curves show the normal-
fluid components based on the measurements above T_λ (see [19] for details).
Note that the comparisons of model and experiment in Fig. 15 are not equivalent to those in Fig. 7. In constructing Fig. 7, the superfluid components
which were added to the normal-fluid components (dashed curves) to get the
solid curves, were calculated from the corresponding distributions observed at
the lowest temperature (1.00 K) assuming that the intensity of the one-phonon
peak varied with temperature as ρ_S/ρ and that its intrinsic width was as
given by the Landau-Khalatnikov theory but that its frequency was independent
of temperature. In constructing Fig. 15, a fit (of a Lorentzian convoluted
with the known Gaussian experimental resolution) was made to the sharp one-
phonon peak in the superfluid component to determine the best values of its
intensity, frequency and intrinsic width (see [19] for details). This
explains why the description of the position and intensity of the sharp one-
roton peak is much better in Fig. 15 than in Fig. 7.

The WS model, as applied in constructing Fig. 15, gives an excellent description of the full $S(Q,\omega)$ for all T < T_λ for the maxon wave vector, 1.13
$\text{\AA}^{-1}$. For the roton wave vector, 2.03 $\text{\AA}^{-1}$, we see that the description is not
adequate at low frequencies (< 0.1 THz). Here, for all T < 1.83 K (see Fig. 1
of [19]), the normal-fluid component, $(\rho_N/\rho)S_N(Q,\omega)$, is larger than the
total observed intensity. This implies a negative intensity in this region
for the superfluid component, which is of course not possible. Hence, either
the weight of the normal-fluid component must fall off more rapidly than
ρ_N/ρ as the temperature is lowered below T_λ or $S_N(Q,\omega)$ must change in
shape and/or position more than accounted for by the simple population factor
used in the WS model, or both. While the nature of the "normal-fluid
component" in the WS model is admittedly not at all clear for T < T_λ, the
zero-sound "mode" which corresponds to the broad low frequency peak observed
above T_λ for Q = 2.03 $\text{\AA}^{-1}$ (dashed curve in Fig. 14) is a very poorly defined
mode (FWHM $\approx$ 2.7 $\times$ peak position) and, in so far as the "extension" of this
mode to lower temperatures has any meaning, it might be expected to renorma-
lize quite dramatically as the temperature decreases. Aside from the problem
in the region of the low frequency wing of the one-phonon peak, the WS model
gives a good description of $S(Q,\omega)$ for the roton. While the problem at low
frequencies shows an inadequacy of the model which may cause small systematic
errors in the values of the one-phonon intensity, $Z(Q,T)$, inferred from the
analysis, it has essentially negligible effect on the values of the one-phonon
frequency, $\omega(Q,T)$, and intrinsic width, $\Gamma(Q,T)$.

The one-phonon parameters for the roton at P = 20 bars obtained [19] using
the WS model are shown in Fig. 16. (A complete tabulation of values for both
wave vectors studied is given in [19].) The solid curve shows ρ_S/ρ
arbitrarily normalized to the value of $Z(Q,T)$ at 1.29 K. While one could make
the overall agreement considerably better by a different choice of normali-
zation, the indications are that $Z(Q,T)$ for Q = 2.03 $\text{\AA}^{-1}$ at 20 bars does not
decrease quite as rapidly with increasing temperature as does ρ_S/ρ. This is
also found to be the case for the maxon at P = 20 bars (see Fig. 7 of [19])
and, as we noted in connection with the right half of Fig. 7, was also the
case for the roton at SVP. As we discussed in the previous section, the WS
model implicitly assumes that $S(Q)$ and the density, ρ, are independent of
temperature for T < T_λ. If this is not the case, then one would not expect
the values of $Z(Q,T)$ obtained from the WS analysis to scale as ρ_S/ρ. We
also noted that, at SVP, ρ only varied by $\approx$ 0.7% below T_λ but that, for the
roton wave vector, $S(Q)$ varied by about 5%. The variation of $S(Q)$ and the
significant change in the roton energy with temperature were believed to be
the reason that, in application (b), the WS model gave a considerably poorer
description (see Fig. 7) of $S(Q,\omega)$ for the roton wave vector than for the

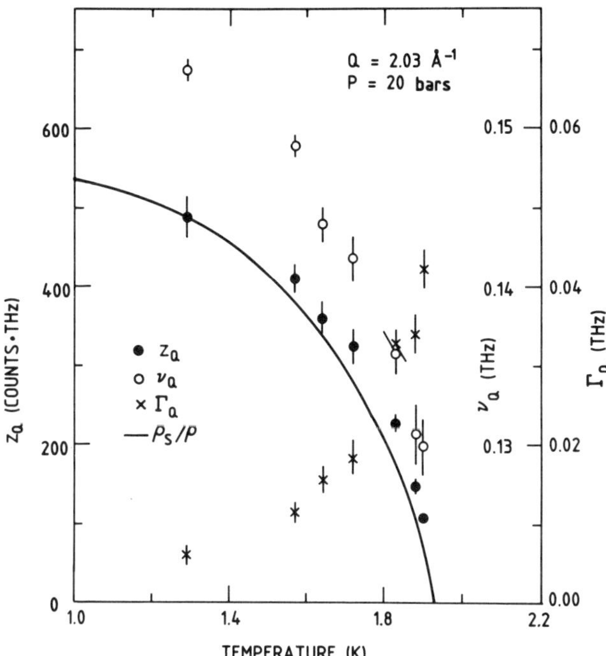

Fig. 16. Temperature dependence of the one-phonon intensities, frequencies and intrinsic widths for rotons in superfluid ^{4}He at P = 20 bars obtained using the WS model. The solid curve shows ρ_S/ρ normalized to the experimental intensity at T = 1.29 K. Results taken from Table I of TALBOT et al. [19]

other wave vectors studied, where S(Q) and ω(Q,T) are relatively insensitive to temperature. The x-ray measurements of WIRTH and HALLOCK [61] show that S(Q) for liquid ^{4}He at a constant density of 0.171 g/cm^3 (close to the density, 0.172 g/cm^3, at T_λ for P = 20 bars [41]) increases by about 5% between 1.16 K and T_λ for the roton wave vector but, as at SVP, is largely independent of temperature for the maxon wave vector. One would expect a somewhat larger increase at constant pressure, as was used for the study [18,19] we are discussing. Also, for P = 20 bars, ρ increases by about 2% between 1.0 K and T_λ, about three times more than at SVP. Considering these larger variations at the higher pressure, and also the fact that the roton energy is lower and more temperature dependent (Fig. 17), it is not at all surprising that the Z(Q,T) values obtained from the WS analysis appear not to scale as closely with ρ_S/ρ at 20 bars as at SVP. (Note that the uncertainty of ± 0.02 K in the experimental temperatures can also account for essentially all of the deviation from the ρ_S/ρ curve at the highest temperature.) Complicating factors of sufficient importance have been ignored in the analysis (both at SVP and 20 bars) that one should not attempt to ascribe any great significance to the precise temperature variation of the Z(Q,T) values obtained. The one conclusion we can make is that, for all Q values studied at both SVP and 20 bars, the intensity of the sharp feature in S(Q,ω), which we have associated with one-phonon excitations, falls off rapidly with increasing temperature and appears to go to zero at T_λ. We only need to look at the observed distributions (Figs. 4, 13 and 14) to reach this conclusion. The WS

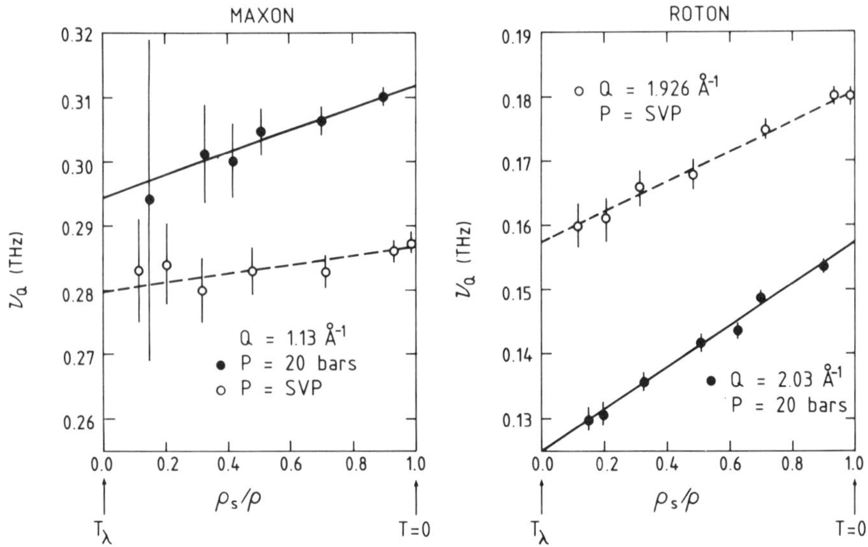

Fig. 17. The one-phonon frequencies for maxons and rotons in superfluid ^{4}He at SVP and P = 20 bars plotted as a function of ρ_S/ρ. These values were obtained by analysis in terms of the WS model and are taken from Table I of TALBOT et al. [19] (for P = 20 bars) and from WOODS and SVENSSON [15,42] (for SVP). The solid and dashed curves are linear fits done by eye.

model, as written down in (1), anticipates that the one-phonon intensity will vary as ρ_S/ρ, and the analysis to date using application (a) of this model indicates that this is indeed roughly the case (Figs. 6 and 16). To check more closely would require a more sophisticated analysis which takes into account all of the complicating factors that we have identified above but that have thus far been ignored. As I stated earlier, this hardly seems to be merited for a purely empirical model. Fortunately, these complicating factors have much less effect on the one-phonon frequencies and widths than on the intensities so we can, I believe, treat the values of $\omega(Q,T)$ and $\Gamma(Q,T)$ as being considerably more firmly established than those of $Z(Q,T)$. We see from Fig. 16 that the roton frequency at P = 20 bars decreases from 0.154 THz at 1.29 K to 0.130 THz at 1.90 K while the width increases from 0.006 THz to 0.042 THz. Note that the roton width at 1.29 K at P = 20 bars is about 4 times larger than the roton width at the same temperature at SVP (see Fig. 8). It is probably more meaningful to compare at the same value (0.67) of the reduced temperature, T/T_λ, but even then the width is a factor of 2 larger at 20 bars.

The one-phonon frequencies for both rotons and maxons at P = 20 bars [19] are compared in Fig. 17 with the values of WOODS and SVENSSON [15,42] at SVP. Here we have plotted the frequencies against ρ_S/ρ and we see that, for both wave vectors at both SVP and 20 bars, they appear to vary linearly with ρ_S/ρ over the full range of temperatures studied. Note that this implies a linear dependence on ρ_N/ρ which in turn means that, in the temperature region studied, where rotons are the dominant thermal excitations and hence make the dominant contribution to ρ_N/ρ, the frequency shifts are essentially proportional to the population of thermal rotons, a not surprising result. The same

84

linear dependence might possibly not continue to very low temperatures where
low-Q phonons become the dominant thermal excitations but on the ρ_S/ρ scale
of Fig. 17 this is a very small region which we can ignore for purposes of the
present discussion. If we use the solid curve through the roton results for
P = 20 bars in Fig. 17 to estimate the total frequency shift between T = 0 and
T = T_λ, we obtain 0.0325 THz (1.56 K). This value falls within the bound,
$\Delta(0) - \Delta(T) < T$, required by the theory of BEDELL et al. [44]. As we have
already noted, this bound is also satisfied by the WS shifts at SVP (see Figs.
8 and 9). Note that the total roton shift, 0.0325 THz (21% decrease in
frequency), at P = 20 bars is considerably larger than the value, 0.0232 THz,
(13% decrease), at SVP. These shifts are, however, much smaller than the
values (40 - 60% decrease at SVP and 70% at P = 20 bars) inferred from earlier
studies [6,23,24,26,28] where the whole observed distributions were analysed,
in some cases, as we have noted above, in terms of inappropriate models. We
see that the maxon frequency decreases much less with increasing temperature
than the roton frequency. The total change for the maxon is only about 2% at
SVP and 5% at P = 20 bars.

The one-phonon widths for maxons and rotons at P = 20 bars [19], are
compared in Fig. 18 with the widths for rotons at SVP [15,42] and the widths
predicted on the basis of the Landau-Khalatnikov theory. Here, as in Fig. 17,
we have plotted against ρ_S/ρ (we might better have chosen ρ_N/ρ but the
reader can easily visualize this alternate plot) to emphasize the dependence
on the population of thermal rotons. (For a plot against temperature see
Fig. 10 in [19].) When plotted this way, the results (both theory and experiment) for rotons at SVP and 20 bars fall very close together suggesting almost
a universal behavior. With the possible exception of the value at the lowest
temperature for P = 20 bars, which we will return to later, there is excellent
agreement between the two sets of experimental results for the roton and
between these results and the LK curves over the whole range of temperatures
studied. We recall (see Fig. 6) that at SVP the widths for maxons and rotons
(as well as several other wave vectors) fall on a universal curve. In
contrast, we see that at P = 20 bars the maxon widths are typically 2-3 times
larger than the roton widths. This is undoubtedly related to the fact that,
at this pressure, the maxon energy is greater than twice the roton energy (see
Fig. 12). Note especially that at the lowest temperature (1.29 K) studied at
P = 20 bars the maxon width is still twice the roton width and that both
widths lie above the LK curve. In contrast to the results at SVP, the widths
at low temperature at P = 20 bars do not scale with ρ_N/ρ, i.e. with the density of thermal rotons. At the higher pressure, some decay process other than
the usual 4-phonon process appears to become important for rotons at low
temperature and to be dominant for maxons at low if not all temperatures. It
is interesting to note that, although the maxon energy exceeds 2Δ for all
temperatures studied [19] at P = 20 bars, if the linear extrapolations to
lower temperatures shown by the solid curves in Fig. 17 are valid, there will
be a crossover at lower temperature ($\approx$ 1.2 K) and the maxon energy will then
lie below 2Δ. It would be very valuable to have high resolution measurements
of the widths and frequencies for rotons and maxons at low temperatures at
P = 20 bars (and probably other pressures) similar to those of MEZEI [29] for
rotons at SVP (see Fig. 8).

DIETRICH et al. [26] have studied the temperature dependence of rotons in
liquid ^{4}He at several pressures. Unfortunately, as we have noted earlier,
they analysed their results (the full observed distributions) in terms of an
inappropriate $\omega \times$ Lorentzian lineshape function. This procedure can only give
reliable values for the one-phonon widths and frequencies when the one-phonon
peak is very sharp and completely dominates the scattering, i.e. at low
temperatures. Not surprisingly, their values for the roton frequencies at a

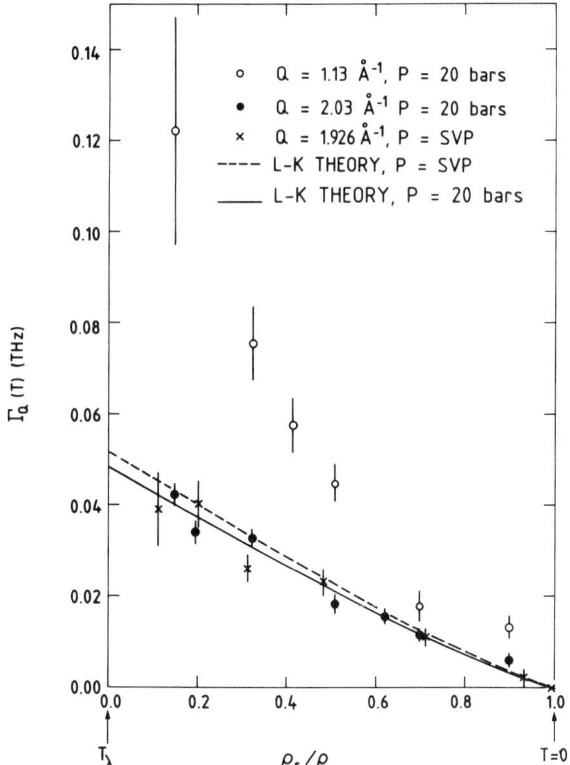

Fig. 18. The one-phonon intrinsic widths for maxons and rotons in superfluid ^{4}He at P = 20 bars, and for rotons at SVP, plotted as a function of ρ_S/ρ. These values were obtained by analysis in terms of the WS model and are taken from Table I of TALBOT et al. [19] (for P = 20 bars) and from WOODS and SVENSSON [15,42] (for SVP). The solid and dashed curves show the widths calculated from the theory of LANDAU and KHALATNIKOV [32] for P = 20 bars and SVP, respectively.

pressure of about 20.3 bars decrease much more with increasing temperature than do those in Fig. 17 (roughly a 70% decrease below T_λ as opposed to only a 21% decrease in Fig. 17) and their width values are considerably larger than those in Fig. 18. BEDELL et al. [45] have calculated roton widths and frequency shifts for a pressure of 24.26 bars for comparison with the highest pressure results of Dietrich et al. The agreement (see Fig. 17 in [45]) is very poor, with the values of Dietrich et al. lying well above the calculated shifts and widths and also violating the bound on the shift derived from roton liquid theory. The pressure (24.26 bars) used in the calculations of Bedell et al. is too different from the pressure (20 bars) of the measurements to allow a meaningful direct comparison with the results shown in Figs. 16-18 but it is clear (see Fig. 11 in [19]) that their theory will give rather good agreement with these results. It would be very worthwhile to have calculations specifically for a pressure of 20 bars.

TALBOT et al. [19] have also analysed their maxon and roton distributions
for P = 20 bars in terms of a different model for the lineshape suggested by
GRIFFIN [55]. In this model, which Talbot et al. refer to as the simple sub-
traction (SS) model, one first separates the distributions observed at the
lowest temperature (i.e., those shown by the solid curves for 1.29 K in Figs.
13 and 14) into one-phonon and multiphonon parts. This separation can be
done, for example, by simply extending the broad multiphonon component
smoothly to zero at low frequency (as was done by Talbot et al.) or, more
systematically [20,21], by making the assumption that the one-phonon peak is a
symmetric function as indicated by the dashed curves in Fig. 2. The separation is clearly less ambiguous for the roton than for the maxon where there is
more overlap between the two components. With the "multiphonon" component,
$S_M(Q,\omega)$, thus determined at the lowest temperature, one then postulates that
it does not change with temperature and subtracts it from the distributions
observed at all higher temperatures, including temperatures above T_λ. The
remainder is treated as the "one-phonon" component and analysed accordingly to
obtain values of the one-phonon parameters, $Z(Q,T)$, $\omega(Q,T)$ and $\Gamma(Q,T)$.

The SS model is a two-component model for $S(Q,\omega)$ at all temperatures, with
one of the components, the "multiphonon component", independent of temperature. It gives "one-phonon" excitations at all temperatures. In contrast,
the WS model is a one-component model for $S(Q,\omega)$ above T_λ (the normal-fluid
component) and effectively a three-component model for $S(Q,\omega)$ below T_λ (the
normal-fluid component and the one-phonon and multiphonon parts of the superfluid component, all of which are temperature dependent). The WS model gives
"one-phonon" excitations only below T_λ. We have already noted that $S(Q,\omega)$
for P = 20 bars is temperature independent above about 0.7 THz (see Fig. 13).
Subtracting a temperature independent $S_M(Q,\omega)$, as in the SS model, will thus
cause the remainder, the supposed "one-phonon" component, to have a tail
extending to but cutting off at $\approx$ 0.7 THz for all temperatures above the
lowest reference temperature, regardless of whether we are dealing with the
maxon or the roton which, at low T, have frequencies which differ by a factor
of two. This does not seem to me to be very reasonable. It also does not
seem at all reasonable to me to take the single broad peak observed above
T_λ, shown by the short-dashed curve in Fig. 13, subtract off the "multiphonon" contribution shown by the solid curve (extended appropriately under
the sharp peak) and treat what is left (a narrow slice along the left side of
the distribution observed at 2.05 K) as the "one-phonon excitation" at that
temperature?

As the reader may have already realized, subtracting a constant $S_M(Q,\omega)$
will, since $S(Q)$ is temperature independent to within about 5%, force the
"one-phonon" intensity to be essentially independent of temperature as is in
fact found [19] for the SS model. My contention is that the experiments (see
Figs. 4, 13, and 14) tell us clearly that the one-phonon intensity, if this is
what we are seeing in the sharp peak as I think it is, is not independent of
temperature but rather falls off rapidly with increasing temperature. This
being the case, I would expect the multiphonon scattering to also change markedly in intensity as the temperature is increased, and probably also somewhat
in shape. A final telling point against the assumption of a temperature independent multiphonon component is that in the only case where we have hard
information, namely at small Q, this assumption is definitely not valid. In
Fig. 19 we show $S(Q,\omega)$ for $Q = 0.3$ Å^{-1} and SVP at 1.2 K and 2.3 K [20]. At
1.2 K the multiphonon component is a well defined peak centred at a frequency
just above 2Δ and completely separated from the one-phonon peak. Two-roton
scattering clearly gives the dominant contribution to $S_M(Q,\omega)$ at this wave
vector. When we go above T_λ, where the roton excitations are either badly
washed out or non-existent, the scattering in the frequency region of the

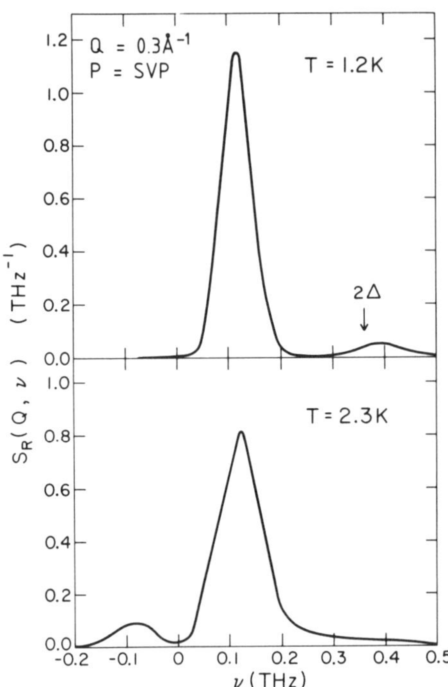

Fig. 19. Resolution-broadened dynamic structure factors of liquid ^{4}He for Q = 0.3 Å^{-1} and SVP at 1.2 and 2.3 K.   From SVENSSON et al. [20]

multiphonon peak at low T changes dramatically.   There is no longer any sign of a peak near 0.4 THz and, even though the main peak has now developed a tail extending into this region, the intensity near 0.4 THz has decreased by more than a factor of two.   We eventually must of course reach the frequency where $S(Q,\omega)$ is independent of temperature but this clearly doesn't occur until close to 0.5 THz or possibly higher.   Since the multiphonon scattering at small Q changes totally in going from 1.2 K to 2.3 K, there is, in my opinion, no reason to believe that the multiphonon scattering should be independent of temperature (as assumed in the SS model [19,55]) for any wave vector.

Since I am not convinced of the merit of the SS model, I am not going to show here the results (see Table II and Figs. 8 and 9 in [19]) from the analysis in terms of this model but I will briefly summarize what is found.   The SS and WS models give the same values for the one-phonon parameters at the lowest temperature, but the values begin to diverge rapidly as the temperature increases.   The intensities, $Z(Q,T)$, which fall off rapidly and roughly as ρ_S/ρ in the WS model, either stay constant or rise substantially near T_λ (depending on interpretation) in the SS model (see Fig. 8 in [19]).   The frequencies, $\omega(Q,T)$, agree for the two models up to about 1.7 K but then the SS values drop much faster than the WS values and, for the roton, the SS frequency shifts are more than a factor of 2 larger near T_λ.   The SS $\omega(Q,T)$ values for rotons are thus in serious disagreement with the values inferred from thermodynamic measurements and also violate the"roton liquid theory" bound on the frequency shift (see Fig. 11 in [19]).   The SS $\Gamma(Q,T)$ values are larger than the WS values for all temperatures above 1.6 K for the roton. Near T_λ, the SS widths for the roton are 2-3 times larger than the WS and

Landau-Khalatnikov widths. The SS model also gives one-phonon parameters
above T_λ with, as in the early studies [6,23,24,26], there being a marked
change in slope in the $\omega(Q,T)$ and $\Gamma(Q,T)$ values at or near T_λ, a simple
consequence of the fact that $S(Q,\omega)$ changes rapidly just below T_λ but
relatively little above (Figs. 13 and 14).

Although we are primarily concerned with neutron scattering results in this
paper, I want to call attention to the fact that OHBAYASHI and his collaborators [62,63] have obtained a very beautiful set of Raman scattering spectra
for liquid ^{4}He, covering a wide range of pressures and temperatures. Their
results for a given pressure show that the roton energy is much less sensitive
to temperature than suggested by DIETRICH et al. [26] and, although a detailed
comparison has not been made, appear to be consistent with the values (Fig.
17) obtained using the WS model. OHBAYASHI [63] has analysed his results for
a pressure of 1 kg/cm^2 ($\approx$ 1 bar) in terms of a two-fluid model analogous to
the WS model. He finds that the Raman spectra for all $T < T_\lambda$ can be rather
well described as an appropriately weighted combination of the "superfluid"
spectrum observed at the lowest temperature (1.3 K), broadened by convolution
with a Lorentzian function whose width is chosen to give a good fit to the low
energy side of the two-roton peak for the temperature under consideration, and
the "normal-fluid" spectrum observed at $T = 2.73$ K. The weights which give
the best description are close to but not precisely ρ_S/ρ and ρ_N/ρ. This
may partly reflect the facts that the temperature (2.73 K) at which the
normal-fluid spectrum was determined was rather far above T_λ ($T_\lambda = 2.16$ K
at $P = 1$ bar) and that the Raman spectra are somewhat more sensitive to
temperature above T_λ than the neutron spectra (at least those for the
relatively large Q values we have been considering in this paper). Note that
the Raman spectra at the higher temperatures in the superfluid phase cannot be
adequately described by a simple thermal broadening of the spectrum observed
at the lowest temperature; an appropriately weighted component of the "shape"
observed above T_λ must be added to get a good description. A two-fluid
model for the scattering thus appears to work very well as an empirical description for both Raman and neutron results even though the quantity measured
by the two methods is quite different. We should also note here that the
strong temperature dependence observed for the Raman spectra (which result
from scattering by combinations of excitations having net $Q \approx 0$ and are
dominated by two-roton processes at low temperatures) also weighs against the
assumption of a temperature independent multiphonon component in the neutron
scattering, as invoked in the SS model.

6. Summary, Conclusions and Future Directions

We have, I believe, made a great deal of progress over the last decade in un-
ravelling the mysteries of liquid ^{4}He. This progress began with a detailed
study [15,17] of the temperature dependence of the neutron scattering at SVP.
This study gave results that were closely proportional to the resolution
broadened dynamic structure factor, $S(Q,\omega)$, and showed, for the first time,
that there was a marked qualitative difference in $S(Q,\omega)$ above and below T_λ.
For all $T < T_\lambda$ there was a sharp peak in $S(Q,\omega)$ but there was no evidence of
such a peak above T_λ (see Fig. 4). This sharp peak, which was the continua-
tion to higher temperatures of the dominant one-phonon peak observed at low
temperature, thus seemed to be a signature of the presence of the Bose conden-
sate/superfluid. We now also have a very similar study [18,19] at a pressure
of 20 bars which shows equally clearly the marked difference in $S(Q,\omega)$ above
and below T_λ (see Figs. 13 and 14). To round out the experimental situation
we also have the really excellent studies of MEZEI [29] for rotons at temperatures below 1.4 K (see Fig. 8) and of MEZEI and STIRLING [30] for low-Q excitations at temperatures below 1.7 K (see Fig. 10). Ideally, it would be nice

to have even more detailed experimental information on what is happening to
$S(Q,\omega)$ close to T_λ. Unfortunately, this would be very difficult and time
consuming to obtain. It is always very hard to study something, like the
sharp peak in $S(Q,\omega)$, that is "disappearing" especially when that something is
sitting on top of a large "background" that is not well understood either. We
can probably never do experiments well enough to satisfy those skeptics who
say "well maybe there is a big change in $\omega(Q,T)$ and $\Gamma(Q,T)$ in the last milli-
degree before T_λ", and, at present, I don't see that there is any pressing
need to try. We already have, in total, a considerable wealth of high quality
experimental results, probably enough to guide us to a true fundamental understanding of liquid ^{4}He if we can just interpret them correctly.

To interpret the experiments we need models and theories, preferably
theories that can give detailed numerical predictions. To date, most of the
effort has been devoted toward obtaining values of the one-phonon parameters
(especially the one-roton parameters) from the experimental results. Early
on, the whole observed distributions were analysed as if they corresponded to
one-phonon scattering but it has been clear for quite some time that this is a
totally unreliable procedure. $S(Q,\omega)$ for $T < T_\lambda$ contains more than one
component and some means of separating it into its different components must
be found before a meaningful analysis can be carried out. On the basis of
their measurements at SVP, WOODS and SVENSSON [15] proposed an empirical
two-fluid model (the WS model) for $S(Q,\omega)$. This model allows one to
systematically extract the sharp-peak component from $S(Q,\omega)$ for $T < T_\lambda$ and,
assuming that this component corresponds to "one-phonon" excitations, to
obtain the one-phonon parameters. We have seen that the values of the
frequencies and intrinsic widths thus obtained for rotons are, for all
$T < T_\lambda$ at both SVP and $P = 20$ bars, in very good agreement with the values
inferred from thermodynamic measurements and calculated from the theories of
LANDAU and KHALATNIKOV [32] and BEDELL et al. [44,45] with the parameters
fixed by thermodynamic data. Previous discrepancies of as much as a factor of
3 between the results of neutron and thermodynamic measurements have been
totally eliminated by use of the WS model. We have also noted that the
frequencies and widths obtained using the WS model indicate that the one-
phonon excitations remain rather well defined excitations right up to the
point where they "disappear". The excitation frequencies decrease only
moderately, and essentially linearly with ρ_N/ρ (i.e., the density of thermal
rotons), throughout the superfluid phase (see Fig. 17). There is no indication of a soft-mode behavior at T_λ as has been proposed by RUVALDS [64] and
as is common behavior for excitations as one approaches a phase transition
characterized by a broken symmetry. It almost appears as if these "excitations" continue to exist above T_λ but then no longer have any weight in
$S(Q,\omega)$, i.e. are no longer density fluctuations.

Thus far, the only other model that has been used to separate $S(Q,\omega)$ into
different components at higher temperatures is the SS Model [19,55]. So many
facts rule against the basic tenet of this model, that the multiphonon scattering is independent of temperature, that I cannot take seriously the values
of the one-phonon parameters that it gives except at temperatures far below
T_λ. Far below T_λ there is of course no problem in determining the one-
phonon parameters. As long as the one-phonon peak remains reasonably sharp
and dominates the scattering, as it does at the roton wave vector to quite
high temperatures, we can determine its position and width with little uncertainty. It is when we get within about 0.3 K of T_λ that we run into difficulties and the one-phonon parameters become strongly model dependent. One
could of course just give up at this point saying that things are too complicated to sort out nearer to T_λ. This would be, I feel, an unnecessarily
pessimistic stance that would in fact prevent us from ever really understanding liquid ^{4}He. To claim a fundamental understanding we have to under-

stand in detail how the presence of the Bose condensate, with its accompanying
broken symmetry, affects the dynamics in the superfluid phase, and the clues
to this clearly lie in what happens to $S(Q,\omega)$ close to T_λ. What happens far
away from T_λ just helps to point us in the right direction.

If, at the present time, we want to dissect $S(Q,\omega)$ near T_λ to get at its
component parts then, in my opinion, we have only the WS model. It is a
purely empirical model and it certainly isn't perfect, as evidenced by the
poor description it gives at low frequencies for the roton wave vector at high
pressure (see Fig. 15), but it is the best we have. It is certainly somewhat
unsatisfactory that we really don't know what meaning to assign to the
"normal-fluid" component of the WS model below T_λ. We simply postulate that
as long as there is normal fluid present it continues to scatter neutrons in
essentially the same way as it does above T_λ. The shape only changes as
needed to satisfy detailed balance but, in the spirit of two-fluid models, the
weight is proportional to the amount of normal fluid present, i.e. ρ_N/ρ. In
so far as this approximation has any meaning, it would be expected to be best
close to T_λ, which is just where we need it most. For the roton wave vector
at high pressure, where the scattering above T_λ is peaked at very low
frequency (Fig. 14), this approximation certainly doesn't work to low temperatures (where however the breakdown has little effect on the one-phonon parameters) but even here the problem of negative superfluid intensities goes away
above 1.83 K (see Fig. 1 of [19]), i.e. within 0.1 K of T_λ. Until we have
something better, I am prepared to tentatively accept, for temperatures
essentially right up to T_λ, the values of the one-phonon frequencies,
$\omega(Q,T)$, and intrinsic widths, $\Gamma(Q,T)$, given by the WS model. I find it hard
to believe that the very good agreement, for all $T < T_\lambda$, of these values
with the calculations based on the thoeries of Landau and Khalatnikov and
Bedell et al. (Figs. 8, 9, and 18), and also the linear dependence of the
$\omega(Q,T)$ values on ρ_N/ρ (Fig. 17), can be entirely fortuitous.

Too much emphasis has probably been placed on determining the one-phonon
(especially one-roton) parameters, largely for historical reasons. There is a
lot more to $S(Q,\omega)$ than just the one-phonon peak and we really need to understand all of $S(Q,\omega)$ before we can be confident that we understand any one
part. We must now move to the level where we compare the full observed
distributions for $S(Q,\omega)$ with the corresponding full theoretical spectra,
convoluted of course with the known experimental resolution. First we need
theories in a sufficiently advanced form that they can be used to make
detailed numerical predictions. As yet we don't have this situation, at least
for the temperature dependence of $S(Q,\omega)$ at the relatively large wave vectors
of principal interest in this paper. This is not meant to imply that there
has not been substantial progress on the theoretical front. In fact there has
been some exceedingly important progress. The theoretical work of GRIFFIN and
his collaborators [53,55,56] has clearly demonstrated that the Bose broken
symmetry plays a crucial role in determining the dynamics of superfluid ^{4}He.
As we noted earlier, their microscopic theory can account, at least qualitatively, for a marked difference in $S(Q,\omega)$ above and below T_λ, as observed in
the neutron scattering experiments, but as yet detailed numerical calculations
based on this theory have not been carried out. We are badly in need of such
calculations.

The theoretical calculations of TALBOT and GRIFFIN [53] for $Q = 0.35$ Å^{-1}
and $Q = 0.8$ Å^{-1} indicate that $S(Q,\omega)$ has just a single peak, centred at the
one-phonon frequency, at all temperatures. Except for the rather weak
multiphonon peak near 2Δ at 1.2 K and the high frequency tail at 2.3 K, this
is essentially what is observed at $Q = 0.3$ Å^{-1} (Fig. 19). Here the single
strong (zero-sound) peak observed above T_λ (the negative energy peak is just

energy-gain scattering from the same mode) is centred at very nearly the same frequency as the one-phonon peak observed at 1.2 K. Under the rather poor resolution conditions used [20] for these measurements (necessary to achieve an energy transfer of 0.5 Thz at such small Q) the widths of the peaks at 1.2 and 2.3 K are rather similar but there is in fact a large difference in intrinsic width. We know from other work [21,30] that the intrinsic width at 1.2 K is very small and that the width increases steadily (see Fig. 10), while the energy changes very little, as the temperature rises in the superfluid phase. Although the actual widths calculated by TALBOT and GRIFFIN [53] are considerably smaller than those obtained from the experiments [15,21,30], the structure of their theory is such that it should be able to explain the temperature dependence of $S(Q,\omega)$ at small Q rather well. As we have noted earlier, it would be very valuable to have results at small Q, of the quality of those of MEZEI and STIRLING [30], extending to temperatures above T_λ to provide a detailed test of the theory of Talbot and Griffin. The approximations these authors use are such that their calculations are only strictly valid at sufficiently small Q that there is negligible multiphonon scattering. We note that this is nearly the case at $Q = 0.3$ Å^{-1} (Fig. 19). It is very important that this theory, which is in principle capable of explaining the temperature dependence of $S(Q,\omega)$ at any Q, be expanded to properly include multiphonon effects so it can be used for detailed numerical calculations at the maxon and roton wave vectors, where $S(Q,\omega)$ is more complicated than at low Q, and hence more interesting, and where we already have sets of top quality experimental results at both SVP and 20 bars. What the experiments seem to show here, a sharp peak popping out of a broad component (often at a frequency well away from the centre of the broad component) as we go below T_λ and then growing rapidly while the broad component slowly fades away (see Figs. 4, 13 and 14), is quite different from what the theory in its present form predicts and what is in fact observed at low Q. One of the aims in proposing [55] the SS model was clearly to try and rid the experimental distributions of the multiphonon scattering so that what was left could be compared with a theory [53] which didn't include multiphonon effects. This approach could work if we really understood the multiphonon scattering and its temperature dependence, which we don't, but what was actually done [19,55], i.e. to "remove" the multiphonon scattering by assuming that it was independent of temperature, was, in my opinion, totally unsatisfactory. What we really have to do is to build multiphonon effects properly into the theory so that we can make direct comparisons with the full experimental distibutions.

If we have to build multiphonon effects into the finite-temperature microscopic theory of TALBOT and GRIFFIN [53], we are first going to have to achieve a better understanding of the multiphonon scattering at low temperature. The measurements of SVENSSON et al. [20] (see Fig. 2), GRAF et al. [37] and STIRLING [65] have shown that the multiphonon scattering, $S_M(Q,\omega)$, has definite structure. From the theoretical studies of JACKSON [66], IWAMOTO et al. [67], GÖTZE and LÜCKE [68], MANOUSAKIS and PANDHARIPANDE [69] and FUKUSHIMA et al. [70] we know that there should be such structure and that, in particular, one should be able to identify various features as arising from roton-roton, roton-maxon and maxon-maxon processes. The work of MANOUSAKIS and PANDHARIPANDE [69] is the most complete to date, but we still need additional theoretical studies, and possibly more measurements (though STIRLING [71] has some even more impressive results, as yet unpublished), in order to advance our understanding to the level where we achieve quantitative agreement between theory and experiment for the complete $S_M(Q,\omega)$ as well as, on the same absolute scale, the one-phonon scattering. This work should be a priority for the near future. One should also mention here that the Raman scattering results of OHBAYASHI and collaborators [62,63] show numerous structural features that are undoubtedly related to combinations of maxons and rotons. A correct interpretation of these features may also prove valuable for understanding the neutron scattering results, or vice versa. More effort should be made to correlate the results of Raman scattering and neutron scattering experiments than has generally been done in the past. We need to maintain as broad a perspective as possible in our quest to understand liquid ^{4}He. This includes taking note of the fact that $S(Q,\omega)$ for non-superfluid ^{4}He is very similar to the zero-sound part of $S(Q,\omega)$ for liquid ^{3}He even at extremely low temperatures like 40 mK [72], and also bears considerable resemblance to the phonon scattering by solid ^{4}He (see [19] for discussion). It is superfluid ^{4}He that is unique in the sense of having exceedingly sharp one-phonon excitations at low temperature and an $S(Q,\omega)$ which exhibits a very strong temperature dependence, undoubtedly consequences of the presence and changing magnitude of the Bose condensate.

To give a final summation, we are at the stage where we have experiments [15,17-19] which show a marked difference in $S(Q,\omega)$ above and below T_λ and a theory [53,55,56] which says that this should be expected. Both show that the condensate plays a crucial role in determining the dynamics of superfluid ^{4}He. The realization that it was of the utmost importance to do measurements near T_λ and considerable high quality experimental and theoretical work over the past decade have gotten us to this point and we are, I think, poised for a breakthrough in our understanding of liquid ^{4}He. It seems as if the answer must be staring us in the face, but we are not quite able to recognize it. We still don't fully comprehend what is happening to $S(Q,\omega)$ near T_λ but I am optimistic that, by the time of the next conference of this type, we will have done so and can then claim, after more than 80 years of puzzling over liquid ^{4}He, to have finally achieved a true fundamental understanding.

Acknowledgements

I have benefitted from and greatly enjoyed numerous stimulating discussions on the topic of this paper with Drs. H.R. Glyde, A. Griffin, K. Ohbayashi, L. Passell, R. Scherm, W.G. Stirling, E.F. Talbot and A.D.B. Woods. I also wish to thank Drs. K. Ohbayashi and M. Watabe for inviting me to participate in the very enjoyable Hiroshima Symposium and for their warm and generous hospitality during my stay in Hiroshima.

References

1. L. Landau: J. Phys. USSR <u>11</u>, 91 (1947). Even earlier [J. Phys. USSR <u>5</u>, 71 (1941)], Landau had postulated the existence of rotons but at that time he envisaged them as forming a separate branch of excitations with energy-momentum dispersion relation $\varepsilon = \Delta + p^2/2\mu$ where μ is an effective mass. In 1947 he changed the roton dispersion relation to $\varepsilon = \Delta + (p - p_0)^2/2\mu$ and connected the phonons ($\varepsilon = cp$, where c is the velocity of sound) and rotons to form the continuous dispersion relation shown as an inset in Fig. 1
2. H. Palevsky, K. Otnes, K.E. Larsson, R. Pauli, R. Stedman: Phys. Rev. <u>108</u>, 1346 (1957)
3. J.L. Yarnell, G.P. Arnold, P.J. Bendt, E.C. Kerr: Phys. Rev. Lett. <u>1</u>, 9 (1958) and Phys. Rev. <u>113</u>, 1379 (1959)
4. H. Palevsky, K. Otnes, K.E. Larsson: Phys. Rev. <u>112</u>, 11 (1958)
5. D.G. Henshaw: Phys. Rev. Lett. <u>1</u>, 127 (1958)
6. K.E. Larsson, K. Otnes: Ark. Fys. 15, 49 (1959)
7. M. Cohen, R.P. Feynman: Phys. Rev. <u>107</u>, 13 (1957)
8. A.D.B. Woods, R.A. Cowley: Rep. Prog. Phys. <u>36</u>, 1135 (1973)

9. D.L. Price: In The Physics of Liquid and Solid Helium, Part II, ed. by
   K.H. Bennemann, J.B. Ketterson (Wiley, New York 1978) p.675
10. R.A. Cowley: In Quantum Liquids, ed. by J. Ruvalds, T. Regge
   (North-Holland, Amsterdam 1978) p.27
11. H.R. Glyde: In Condensed Matter Research Using Neutrons, ed. by S.W.
   Lovesey, R. Scherm (Plenum, New York 1984) p.95
12. E.C. Svensson, V.F. Sears: In Frontiers of Neutron Scattering, ed. by
   R.J. Birgeneau, D.E. Moncton, A. Zeilinger (North-Holland, Amsterdam 1986)
   p.126 [Reprinted from Physica 137B, 126 (1986)]
13. H.R. Glyde, E.C. Svensson: In Neutron Scattering, ed. by D.L. Price,
   K. Sköld, Methods of Exp. Phys., Vol. 23, Part B (Academic Press, New York
   1987) p.303
14. R.J. Donnelly, J.A. Donnelly, R.N. Hills: J. Low Temp. Phys. $\underline{44}$, 471
   (1981)
15. A.D.B. Woods, E.C. Svensson: Phys. Rev. Lett. $\underline{41}$, 974 (1978)
16. B.N. Brockhouse: In Inelastic Scattering of Neutrons in Solids and Liquids
   (International Atomic Energy Agency, Vienna 1961) p.113
17. E.C. Svensson, R. Scherm, A.D.B. Woods: J. de Phys. $\underline{39}$, C6-211 (1978)
18. E.C. Svensson, W.G. Stirling, E. Talbot, H.R. Glyde: In Proc. 18th Int.
   Conf. on Low Temperature Physics, Jpn. J. Appl. Phys. $\underline{26}$ (Suppl. 26-3), 33
   (1987)
19. E.F. Talbot, H.R. Glyde, W.G. Stirling, E.C. Svensson: Phys. Rev. B $\underline{38}$,
   11229 (1988)
20. E.C. Svensson, P. Martel, V.F. Sears, A.D.B. Woods: Can. J. Phys. $\underline{54}$, 2178
   (1976)
21. R.A. Cowley, A.D.B. Woods: Can. J. Phys. $\underline{49}$, 177 (1971)
22. P. Martel, E.C. Svensson, A.D.B. Woods, V.F. Sears, R.A. Cowley: J. Low
   Temp. Phys. $\underline{23}$, 285 (1976)
23. D.G. Henshaw, A.D.B. Woods: In Proc. VIIth Int. Conf. on Low Temperature
   Physics, ed. by G.M. Graham, A.C. Hollis Hallett (Univ. of Toronto Press,
   Toronto 1961) p.539
24. D.G. Henshaw, A.D.B. Woods: Phys. Rev. $\underline{121}$, 1266 (1961)
25. A.D.B. Woods: Phys. Rev. Lett. 14, 355 (1965)
26. O.W. Dietrich, E.H. Graf, C.H. Huang, L. Passell: Phys. Rev. A $\underline{5}$, 1377
   (1972)
27. N.M. Blagoveshchenskii, E.B. Dokukin: Sov. Phys. JETP Lett. $\underline{28}$, 363 (1978)
28. J.A. Tarvin, L. Passell: Phys. Rev. B $\underline{19}$, 1458 (1979)
29. F. Mezei: Phys. Rev. Lett. $\underline{44}$, 1601 (1980)
30. F. Mezei, W.G. Stirling: In 75th Jubilee Conf. on Helium-4, ed. by J.G.M.
   Armitage (World Scientific, Singapore 1983) p.111
31. R. Golub, C. Jewell, P. Ageron, W. Mampe, B. Heckel, I. Kilvington: Z.
   Phys. B $\underline{51}$, 187 (1983)
32. L.D. Landau, I.M. Khalatnikov: Zh. Eksp. Teor. Fiz. $\underline{19}$, 637 (1949)
33. P.J. Bendt, R.D. Cowan, J.L. Yarnell: Phys. Rev. $\underline{113}$, 1386 (1959)
34. J.S. Brooks, R.J. Donnelly: J. Phys. Chem. Ref. Data $\underline{6}$, 51 (1977)
35. A.D.B. Woods, E.C. Svensson, P. Martel: In Neutron Inelastic Scattering
   1972 (International Atomic Energy Agency, Vienna 1972) p.359
36. A.D.B. Woods, E.C. Svensson, P. Martel: Phys. Lett. $\underline{43A}$, 223 (1973)
37. E.H. Graf, V.J. Minkiewicz, H. Bjerrum Møller, L. Passell: Phys. Rev. A
   $\underline{10}$, 1748 (1974)
38. B.A. Dasannacharya, A. Kollmar, T. Springer: Phys. Lett. $\underline{55A}$, 337 (1976)
39. A.D.B. Woods, E.C. Svensson, P. Martel: Can. J. Phys. 56, 302 (1978)
40. A. Miller, D. Pines, P. Nozières: Phys. Rev. $\underline{127}$, 1452 (1962)
41. J. Maynard: Phys. Rev. B $\underline{14}$, 3868 (1976)
42. A.D.B. Woods, E.C. Svensson: Unpublished work based on the study reported
   in [15]
43. V.P. Mineev: Sov. Phys. JETP Lett. $\underline{32}$, 489 (1980)
44. K. Bedell, D. Pines, I. Fomin: J. Low Temp. Phys. $\underline{48}$, 417 (1982)

45. K. Bedell, D. Pines, A. Zawadowski: Phys. Rev. B $\underline{29}$, 102 (1984)
46. E.C. Svensson, V.F. Sears, A.D.B. Woods, P. Martel: Phys. Rev. B $\underline{21}$, 3638 (1980) and V.F. Sears, E.C. Svensson, A.D.B. Woods, P. Martel: Atomic Energy of Canada Limited Report No. AECL-6779 (1979)
47. E.C. Svensson, V.F. Sears, A. Griffin: Phys. Rev. B 23 4493 (1981)
48. E.C. Svensson, A.F. Murray, Physica $\underline{108B}$, 1317 (1981)
49. T.J. Greytak, J. Yan: In Proc. 12th Int. Conf. on Low Temperature Physics, ed. by E. Kanda (Keigaku Publ. Co., Tokyo 1971) p.89
50. A.D.B. Woods, P.A. Hilton, R. Scherm, W.G. Stirling: J. Phys. C $\underline{10}$, L45 (1977)
51. A. Griffin: Phys. Rev. B $\underline{19}$, 5946 (1979), Phys. Lett. $\underline{71A}$, 237 (1979) and J. Low Temp. Phys. $\underline{44}$, 441 (1981)
52. A. Griffin, E. Talbot: Phys. Rev. B 24, 5075 (1981)
53. E. Talbot, A. Griffin: Ann. Phys. (N.Y.) $\underline{151}$, 71 (1983) and Phys. Rev. B $\underline{29}$ 2531 (1984)
54. K. Yamada: Prog. Theor. Phys. $\underline{63}$, 715 (1980) and K. Ishikawa, K. Yamada: These proceedings
55. A. Griffin: Can. J. Phys. $\underline{65}$, 1368 (1987) and these proceedings
56. S.H. Payne, A. Griffin: Phys. Rev. B 32, 7199 (1985) and A. Griffin, S.H. Payne: J. Low Temp. Phys. 64, 155 (1986)
57. H.J. Maris: Rev. Mod. Phys. 49, 341 (1977)
58. W.G. Stirling: In 75th Jubilee Conf. on Helium-4, ed. by J.G.M. Armitage (World Scientific, Singapore 1983) p.109
59. E.C. Svensson, D.C. Tennant: In Proc. 18th Int. Conf. on Low Temperature Physics, Jpn. J. Appl. Phys. 26 (Suppl. 26-3), 31 (1987)
60. V.K. Wong: Phys. Lett. 61A, 454 (1977)
61. F.H. Wirth, R.B. Hallock: Phys. Rev. B 35, 89 (1987)
62. K. Ohbayashi, M. Udagawa: Phys. Rev. B 31, 1324 (1985), M. Udagawa, H. Nakamura, M. Murakami, K. Ohbayashi: Phys. Rev. B 34, 1563 (1986), M. Udagawa, T. Imada, K. Ohbayashi: J. Phys. C 20, 1063 (1987), K. Ohbayashi, M. Udagawa, M. Watabe: Can. J. Phys. 65, 1571 (1987) and M. Udagawa, M. Watabe, K. Ohbayashi: These proceedings
63. K. Ohbayashi: Private communication and these proceedings
64. J. Ruvalds: Phys. Rev. Lett. 27, 1769 (1971)
65. W.G. Stirling: In Proc. 2nd Int. Conf. on Phonon Physics, ed. by J. Kollar, N. Kroo, N. Menyhard, T. Siklos (World Scientific, Singapore 1985) p.829
66. H.W. Jackson: Phys. Rev. A 4, 2386 (1971) and 8, 1529 (1973)
67. F. Iwamoto, K. Nagai, K. Nojima: In Proc. 12th Int. Conf. on Low Temperature Physics, ed. by E. Kanda (Keigaku Publ. Co., Tokyo 1971) p.189
68. W. Götze, M. Lücke: Phys. Rev. B 13, 3825 (1976)
69. E. Manousakis, V.R. Pandharipande: Phys. Rev. B 33, 150 (1986) and E. Manousakis: Ph.D. Thesis, Univ. of Illinois at Urbana Champaign, 1985
70. K. Fukushima, N. Koyama, T. Sugiyama: Prog. Theor. Phys. 61, 367 (1979) and K. Fukushima, F. Iseki: Phys. Rev. B 38, 4448 (1988) and these proceedings.
71. W.G. Stirling: Private communication
72. K. Sköld, C.A. Pelizzari: Phil. Trans. R. Soc. Lond. B 290, 605 (1980)

Upscattering of Ultra-Cold Neutrons (UCN) by Excitations in Superfluid ^{4}He

*R. Golub**

Physics Department, Technical University Berlin, Germany

Ultra-Cold Neutrons have energies $\simeq 10^{-7}$ eV, velocities $\simeq$ 5m/sec and wavelength $\simeq$ 500 Å. They rise only about 1 meter in the Earth's gravitational field and can be totally reflected by magnetic fields $\simeq$ 40 kilogauss and material surfaces. A collection of atoms appears as a group of hard sphere repulsive potentials to a neutron. The UCN, with wavelengths $\gg$ interatomic spacing are affected by the space average of these potentials which is about 10^{-7} eV for many materials. On total reflection the UCN wavefunction penetrates into the reflecting surface and can be absorbed or upscattered by the thermally vibrating nucleii located there.

UCN have been stored in material bottles with decay times $\simeq$ 600 seconds, this means that one can detect stored UCN even after they have been in the vessel for one hour – UCN can be stored in liquid He^4 at temperatures $\leqslant$ 1 K.

The upscattering of UCN by Helium^4 is interesting because a) it allows the direct determination of the UCN density inside the Helium – it allows us to look through the walls of the container into the liquid,
b) it allows the study of the 3 phonon interaction in the Helium at Q and T values inaccessible to other methods.

Plotting the energy–momentum relation for a free neutron ($E = P^2/2m$) on the usual He^4 dispersion curve the point of intersection will be designated by E^*, P^*. Only neutrons of energy E^* ($= 11$ K) can be brought to rest by the emission of a single phonon.

Conversely the only excitations which can be absorbed by a neutron at rest are phonons of Energy E^*. If we generalize from neutrons at rest to UCN, we

*Supported by: Bundesministerium für Forschung und Technologie (W.–Germany) and Society for the Promotion of Accelerator Physics (Japan)

Springer Series in Solid-State Sciences, Vol. 79   **Elementary Excitations in Quantum Fluids**
Editors: K. Ohbayashi · M. Watabe   © Springer-Verlag Berlin, Heidelberg 1989

see that the above considerations will apply to a narrow range of energies around E*. Since only phonons in the vicinity of E* can be absorbed by a UCN the upscattering rate $1/\tau_{ip} \simeq e^{-E^*/T}$ and becomes rather small for temperatures $T \leqslant 1$ K. Thus if a vessel with good UCN containing walls and filled with He^4 is exposed to a neutron flux containing neutrons with energy E*, there will be a constant production rate of P UCN/cm^3/sec in the Helium and the UCN density will build up to $\rho_{UCN} = P\tau_0$ where τ_0 is the average storage time due to all loss mechanisms. The dominant He^4 contribution to τ_0 is phonon scattering from the neutron (involving the 3 phonon interaction) but there is a contribution from Roton scattering and a residual contribution from the one phonon absorption.

The UCN density in the Helium can be much greater than would be expected by applying Liouville's theorem to the incident neutron flux alone.

An apparatus containing a 3 meter long, Helium filled, UCN storage chamber has been installed on a cold neutron guide at the ILL and measurements of the temperature dependence of the storage time are in reasonable agreement with calculations but the discrepancies may be due to our lack of knowledge concerning the 3 phonon interaction. The measurements indicate a somewhat weaker interaction than was used by Maris, but measurements of phonon linewidths also seem to indicate a somewhat weaker interaction.

The cryostat was fitted with an exit system consisting of two detectors – one in line with the 3 meter storage chamber and the incident beam for detecting the upscattered neutrons, and the second, located 2 meters away behind a concrete partition and connected to the system by two right angle bends, measured the UCN. The upscattered neutrons were separated from the UCN by a 5mm Be foil placed at 45º to the axis of the apparatus which reflected the UCN towards the UCN detector.

The upscattered neutrons were measured as a function of time while the UCN were being stored (UCN valve closed), while the UCN were measured by opening the UCN valve at varying times during the storage cycle.
The UCN density behaves as

$$\rho = \rho_0 \, e^{-t/\tau(T)} \qquad (1)$$

where $\rho_0 = P\tau$ is the saturation UCN density, t is measured from the time of switching off the incident flux, and $\tau(T)$, the temperature dependent measured decay time is given by

$$1/\tau(T) = 1/\tau_0 + 1/\tau_{up}(T) \quad \text{where} \tag{2}$$

τ_0 is the decay time due to temperature independent losses such a absorption on He^3 and impurities and can be measured at the lowest achievable temperature, and $\tau_{up}(T)$ is the upscattering cross section due to interaction with the He^4. The upscattering rate is given by

$$U = \varrho/\tau_{up}(T) \tag{3}$$

which yields

$$U(t = 0) \equiv U_0 = P \left[1 - \frac{\tau(T)}{\tau_0} \right] \tag{4}$$

using (1) and (2).

The measured upscattering showed the time dependence expected according to (3). In all cases the decay time $\tau(T)$ determined from the upscattering data was in good agreement with that measured directly using the UCN. Plotting the upscattering intercepts, U_0 vs $\tau(T)$ we see that equation (3) is satisfied, the intercept yielding $\tau_0 = 51$ seconds in good agreement with that measured directly using UCN. Note that the temperature dependence according to (4) is dramatically different from that of the UCN density ϱ_0 which is found to follow the expected $\varrho_0 = P \cdot \tau(T)$ relation.

The results indicate a production rate of 1 $UCN/cm^3/sec$ in comparison with 2.2 $UCN/cm^3/sec$ expected on the basis of measurements of the incident flux.

Based on these results we are preparing an experiment to measure the energy spectrum of the upscattered neutrons.

"In the search for truth there are certain questions that are unimportant. Of what material is the universe constructed?...
If a man were to postpone his searching for Enlightenment until such questions were solved, he would die before he found the path."
– The teaching of Buddha,
                Buddhist Promoting Foundation

References

A.I. Kilvington + R. Golub – Physics Letters, to be published

R. Golub, C. Jewell, P. Ageron, W. Mampe, B. Heckel and A.I. Kilvington – Z. Phys. B $\underline{51}$, 187 (1983)

Excitations in ^{4}He Films

H.J. Lauter[1], V.L.P. Frank[1,2], H. Godfrin[1], and P. Leiderer[2]

[1]Institut Laue-Langevin, 156X, F-38042 Grenoble, France
[2]Institut für Physik, J. Gutenberg Universität,
  D-6500 Mainz, Fed. Rep. of Germany

1. Introduction

The behavior of a helium film is determined by two dimensional (2d)
properties and also by the properties of the substrate on which it is
adsorbed. The substrate used in this work was graphite. Two layers adjacent
to the substrate can be solidified due to the van der Waals attraction with
the graphite. The densities of the layers are very different, displaying
through the high compressibility of the helium the dependence of the van der
Waals force on the distance from the surface. But also the in-plane
corrugation of the surface potential influences the structure of first and
second layer [1]. Subsequent layers are found to be liquid. But a layered
structure is still retained [2]. The superfluid fraction of a thin film is
reduced with respect to the bulk ^{4}He [3]. This may be caused by a different
excitation spectrum in the film (a 2d one) with a smaller roton gap. The
excitations found at energies below the bulk roton value [4] are also
visible in the neutron scattering of a thick film and even if the film is
extended to bulk liquid [5]. Thus these excitations have to be localized
near the solid liquid interface. These modes show only little or no
dispersion, they exhibit localized behavior. Only at a momentum transfer
below the one of the bulk maxon an excitation has been detected which shows
dispersion like an acoustic phonon with a reduced sound velocity with
respect to bulk helium [6].

2. Solid Layers

Several experiments have been performed to study the behavior and in
particular the structure (neutron diffraction) of the mono [7] and the
bilayer [8] of ^{4}He on graphite. The phase diagram of a monolayer of ^{4}He on
graphite shows that the corrugation of the surface potential has a big
influence. The ^{4}He atoms lock into the ($\sqrt{3}$ x $\sqrt{3}$) commensurate structure [7]
as do the other quantum gases. The transition to the denser incommensurate
phase displays the subtle differences between the quanticity of the
different gases [9]. In the incommensurate phase the corrugation is still of
influence, visible e.g. in the dependence of the intensity of the first
Bragg peak of the ^{4}He layer on the coverage. The second layer becomes a
solid just when the first atoms start to populate the third layer [1]. The
lowest detected areal density is 0.089 atoms/Å^2 in the second solid layer
whereas the first layer density is at 0.117 atoms/Å^2. For higher coverages
the second layer compresses at least to 0.095 atoms/Å^2 [1,10]. These
densities are very different and the question is whether the two layers can
be considered to be completely independent. A first interference appears in
that the additional van der Waals attraction of the second layer atoms with
the substrate also reduces the distance of the first layer atoms to the
substrate. Thus the corrugation of the surface potential becomes more

important. This is seen in a splitting of the diffraction peak arising from the diffraction of the first layer [1], indicating a distortion of the first layer lattice. This effect is substrate dependent, if ZYX-graphite [11] is used which exhibits a higher coherence length no splitting is seen but a lock-in in a higher order commensurate phase [10]. A second interference is expected in the bilayer structure factor if the second layer atoms are preferentially adsorbed on sites given by the first layer atoms. These sites are established by three close packed first layer atoms. The second layer atoms show an enhanced probability to be adsorbed at these sites [1]. But no complete overstructure could be found of the second layer with respect to the first one. In summary there seems to be a lot of stress and strain in the first and second layer also with respect to the substrate in order to find a configuration which minimizes energy. Thus no single phonon excitations are expected to exist in the bilayer but broadened features due to the complicated structure and to life time effects.

3. Excitations in the Roton Region

The reduced superfluid fraction in a helium film [3] gave rise to search for an excitation spectrum with a smaller roton gap compared to the bulk one. Inelastic neutron scattering experiments concentrated first on the region below the gap and succeeded indeed to detect an excitation at about 0.55 meV [4], which is 0.2 meV below the bulk roton minimum. However this excitation was nearly independent of the wavevector in a limited range around the bulk minimum. Still another excitation was detected at about 0.4 meV. These excitations are situated at the liquid-solid interface [5].

A new study was performed to investigate in more detail the origin of the excitations which are bound to the surface. The experiments have been done at the ILL on the IN3 triple axis spectrometer with a focusing analyzer and a fixed end-energy of 1.22 THz. The graphite substrate was Papyex [11] oriented with the (002) planes parallel to the scattering plane. A series of scans is shown in Fig.1 done with a rather coarse resolution of 0.05 THz (FWHM) at fixed momentum transfer of 1.96 Å^{-1} (see Fig.4 for orientation). A

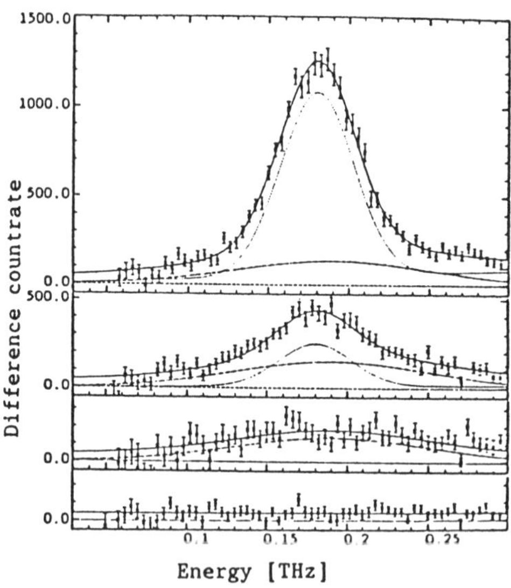

Fig.1 Inelastic neutron scattering scans at constant Q transfer of 1.96 Å^{-1} at a temperature of 0.8 K (the background signal of the sample cell without ^{4}He is subtracted). The different coverages of ^{4}He on graphite are from bottom to top: 2.6, 4.8, 6.26 and 9.8 layers. The solid line is the fit being composed of a constant background (not shown separately), the contribution of the bulk ^{4}He (·····) and the surface signal (–·–).

100

separation of the excitations as mentioned before is not expected and not looked for. A Gaussian line shape was used to fit globally all signal arising from the surface. Another Gaussian sitting on a linear contribution was chosen to account for the bulk roton signal. The linear contribution is due to the finite resolution in connection with the curvature of the dispersion curve and to a small amount to multiple scattering (see next paragraph). This contribution and the parameters of the Gaussian for the roton signal have been fitted at the signal from the 12 layer ^{4}He film sample, since here the surface signal is small compared to the one of the bulk. For the less thicker films the parameters of the bulk signal fit have been kept constant except for the intensity. The intensity originating from the surface is growing with respect to the bulk signal if the film thickness is reduced. This intensity is fitted by another Gaussian as already mentioned. It turns out that the parameters of this Gaussian do not vary as a function of film thickness. This again shows that the properties of this excitation do not depend on the thickness of the film with the restriction that a certain film thickness is required. These features are summarized in Fig.2. A monolayer gives within the statistics no signal. The two and a half layer signal is a flat background which seems to be always present for the thicker films. This is probably due to the complicated equilibrium structure of the bilayer discussed in the preceding paragraph. If the film thickness is further increased the surface signal appears and is the only contribution besides the background up to five layers. Beyond five layers the bulk signal starts to contribute linearly in addition to the constant background and the surface signal. Thus more than three liquid layers are needed above the two solid layers to create a bulk signal in agreement with previous experiments [5], where the threshold was three liquid layers above one solid ^{4}He layer on Ne preplated graphite. In summary the surface signal shows up beyond adsorption of the two solid layers and is independent of the adsorption of further layers proving its origin at the ^{4}He solid-liquid interface. We cannot yet establish a connection to the phonon like mode at smaller momentum transfer,which is analyzed in the next paragraph.

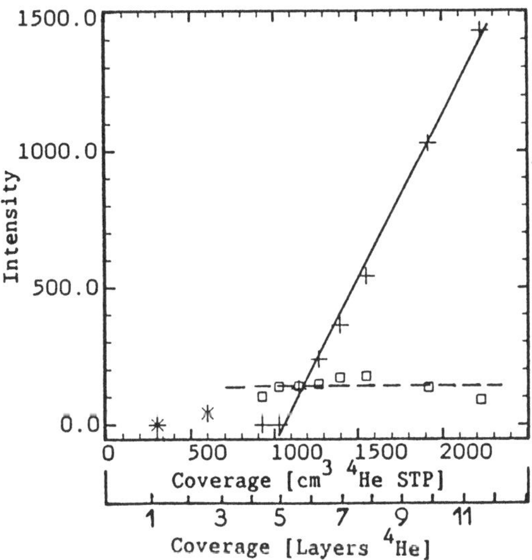

Fig.2 Intensities of different components of the fits in Fig.1 as a function of layer thickness. The signal at a coverage of one monolayer and the one at 2.6 layers (which is a constant background for the subsequent layers) is marked (*), the contribution of the bulk ^{4}He (I) and the surface signal (□).

4. 2-Dimensional Phonon

In another study we focused on excitations at small momentum transfers. The study has been performed on the time of flight spectrometer IN6 at the ILL with an incident wavelength of 5.9 Å. The elastic resolution was about 0.05 meV. The substrate was Vulcan III graphite powder [11]. Some spectra taken at constant scattering angle are shown in Fig.3. The corresponding elastic momentum transfer Q is 1.61 Å^{-1}. In the spectrum from the sample with 7.7 adsorbed layers a high peak is visible at 1.08 meV which is the bulk phonon. This is found again in Fig.4 where the scans taken at constant angle have been transferred into the energy – Q plot. The scan at Q transfer of 1.61 Å^{-1} cuts the bulk dispersion curve at about 1.1 meV, which corresponds to the measured value. This bulk phonon has nearly disappeared for the other coverages of 5 and 3.7 layers in Fig.3 in agreement with the discussion in the preceding paragraph. Another feature in these scans is the multiple scattering. The neutron makes first an elastic scattering at the (002) planes of the graphite and then it creates an excitation in the bulk helium. Because the density of states is very high for the roton region of the dispersion curve the multiple scattering concentrates around this energy transfer. However the multiple scattering is a complicated function of Q and energy transfer [12]. It is represented by the solid line in the figures and is subtracted to leave only the signal from the surface and the bulk. Of course it contributes most at the highest coverage. The signal from the surface consists at this Q transfer of a broad feature ranging from 0.4 meV to 1.2 meV. We concentrate on the structures seen at 0.53, 0.62 and 0.75 meV. It is obvious that the intensity is switching from the higher energy mode to the lower energy one as a function of decreasing coverage. But some signal of each of the three modes is always detectable in the three coverages. This effect is again displayed in Fig.4 to where the location of more pronounced scattering events from other scattering angles were

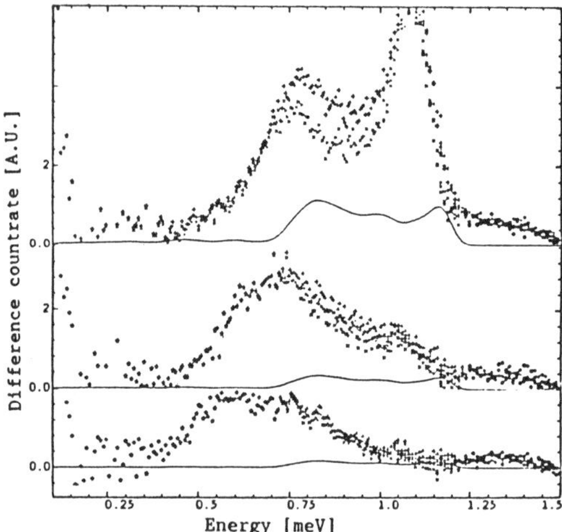

Fig.3 Inelastic neutron scattering scans at constant scattering angle (elastic Q-transfer 1.61 Å^{-1}) at a temperature of 0.72 K for three coverages of 3.7, 5.0 and 7.7 layers (from bottom to top) of ^{4}He on graphite (the background signal of the sample cell without ^{4}He is subtracted). The points marked (x) were obtained by subtracting the multiple scattering (line) from the measured points (+).(1 THz = 4.136 meV)

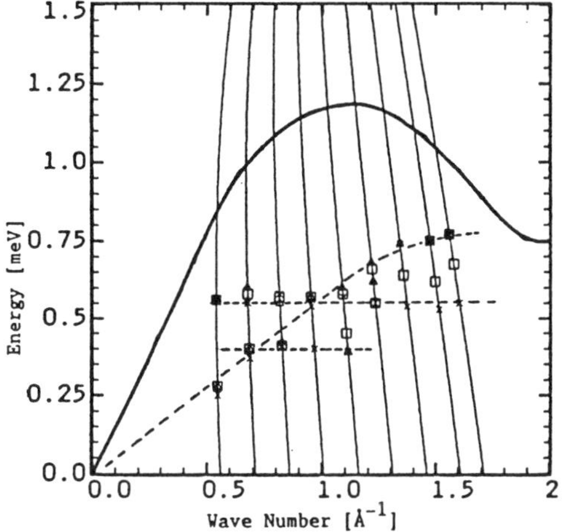

Fig.4 The dispersion curve of bulk ^{4}He is shown as a solid line as well as the constant angle scans (inelastic neutron scattering) on which the different events are marked (see Fig.3).
(x), (□) and (△) are the peak positions of the events of the 7.7, 5.0 and 3.7 layer film, respectively.

transferred. The main features are two modes which are crossing at $Q \approx 1 Å^{-1}$ and $E \approx 0.55$ meV. One is a mode which is constant at 0.55 meV a value also found for surface excitations below the roton [4-6]. The other mode shows a phonon like dispersion. The weight of the intensity is a function of the number of adsorbed layers. For 7.7 layers the main intensity is in the local mode for low Q's up to 1 $Å^{-1}$ and then it follows the higher energy branch of the phonon. For 3.7 layers the main intensity follows first the phonon branch (with a slight interference at 0.4 meV with an other local mode seen also below the roton [5,6]) and then switches to the local mode however still with some contribution to the branches at higher energy (see Fig.3). This behavior is emphasized by the dashed lines in Fig.4. The third coverage at 5 layers is somewhat between the behavior of the two other coverages with a remarkable contribution to the mode at 0.65 meV at higher Q transfer.

In summary there is strong interaction between a phonon like mode and several localized modes and the distribution of intensity is a strong function of coverage. A more detailed analysis will follow [12]. It seems that the localized modes are bound to the solid liquid ^{4}He interface because the interaction becomes stronger as the thickness of the film is reduced. The phonon like mode may have different origins: a pure 2d mode [3], an excitation on a vortex [13] or an excitation of a ripplon [2,14]. A decision cannot yet be made.

5. Acknowledgments

We thank R.J. Donnelly, J. Eckert, H.P. Schildberg and W.F. Vinen for helpful discussions. This work has been partially supported by the Ministry of Research and Technology (BMFT) of the F.R.G..

1. H.J. Lauter, H.P. Schildberg, H. Godfrin, H. Wiechert, R. Haensel: In
   Banff Conf. on Quantum Fluids and Solids, Canad.J.Phys. 65 (Nov.1987)
2. E. Krotscheck: Phys. Rev. B32, 5713 (1985)
   J.L. Epstein, E. Krotscheck to be published (Phys. Rev. B)

3. M. Chester: Comments Solid State Phys. 10, 91 (1981)
   W.F. Saam, M.W. Cole: Phys. Rev. $\underline{B11}$, 1086 (1975)
   W. Götze, M.Lücke: J. Low Temp. Phys. $\underline{25}$, 671 (1976)
4. B. Lambert, D. Salin, J. Joffrin, R. Scherm: J. Physique $\underline{38}$, L377 (1977)
   w. Thomlinson, J.A. Tarvin, L. Passell: Phys. Rev. Lett. $\underline{44}$, 267 (1980)
   H.J. Lauter, H. Wiechert, C. Tiby: Physica $\underline{107B}$, 239 (1981)
5. H.J. Lauter, H. Godfrin, C. Tiby, H. Wiechert, P.E. Obermayer: Surf. Science $\underline{125}$, 265 (1983) and in 75th Jubilee Conference on Helium-4, ed. H.G.M. Armitage (World Scientific, Singapore, 1983) p.84
6. H.J. Lauter, H. Godfrin, H. Wiechert: In Phonon Physics, ed. J.Kollár et.al., 842 (World Science, Singapore 1985)
7. R.E. Ecke, J.G. Dash: Phys. Rev. $\underline{B28}$, 3738 (1983)
   R.E. Ecke, Q.-S. Shu, T.S. Sullivan, O.E. Vilches: Phys.Rev $\underline{B31}$, 448 (1985)
   K. Carneiro, W.D. Ellenson, L. Passell , J.P. McTague , H. Taub: Phys. Rev. Lett. $\underline{37}$, 1695 (1976)
8. S.E. Polanco, M. Bretz: Phys. Rev. $\underline{B17}$, 151 (1978)
   H.J. Lauter, H. Wiechert, R. Feile: In Ordering in Two Dimensions, ed. S.K. Sinha (North Holland, New York 1980) p.291
   K. Carneiro, L. Passell, W. Thomlinson, H.Taub: Phys.Rev. $\underline{B24}$, 1170(1981)
9. H.P. Schildberg, H.J. Lauter, H. Freimuth, H. Wiechert, R. Haensel: In Proceed. LT-18, Japan. J. Appl. Phys. $\underline{26}$, Supp.$\underline{26-3}$, 345 (1987)
10. H.P. Schildberg, H.J. Lauter to be published
11. Papyex is produced by Carbon Lorraine, U.S.A
    ZYX is produced by Union Carbide, France
    Vulcan III is produced by National Physical Laboratory, U.K.
12. V.L.P. Frank, H. Godfrin, H.J. Lauter, P. Leiderer to be published
13. W.I. Glaberson, R.J. Donnelly: In Progress in Low Temperature Physics IX, ed. D.F. Brewer (North Holland, Amsterdam, 1986) p.1
14. D.O. Edwards, W.F. Saam: In Progress in Low Temperature Physics VIIa, ed. D.F. Brewer (North-Holland, Amsterdam,1987) p.283

Part III

Theoretical Study of Elementary Excitations and Related Properties: Raman Scattering and the Dynamic Structure Factor $S(k, \omega)$

Theory of Inelastic Light Scattering from ^{4}He

J.W. Halley

School of Physics and Astronomy, University of Minnesota, Minneapolis, MN 55455, USA

We briefly review the mechanisms of second order light scattering in liquid helium four as well as some key features and results of calculations of the extinction coefficient. We then review the new experimental results of Ohbayashi and his group emphasizing the measurements of the s-d intensity ratio and the observation of fine structure well above twice the roton frequency. Finally we offer some suggestions concerning the theoretical interpretation of the new experimental results.

1. Introduction

Second order inelastic light scattering was first observed in superfluid helium four eighteen years ago in experiments by GREYTAK and YAN [1].   The experiment was preceded by a prediction [2] which, while qualitatively in accord with the experiments, did not get the details right and this provided the occasion for a great deal of interesting theoretical work in the next few years. The most important   new aspect of the subsequent theories was the suggestion [3, 4] that the two rotons excited by the light interact attractively and form a bound   state. Further experiments by the GREYTAK group [5- 6] at higher frequency resolution gave evidence of the expected structure below the frequency corresponding to two free rotons and these experiments were fit quite successfully to the theory of reference 4. With these successes in hand many felt that the fundamental processes involved in this experiment were well understood and interests turned elsewhere. This consensus largely ignored the fact, to be discussed in more detail below, that the theories being used badly violated a fundamental requirement, namely that two helium atoms could not be in the same place at the same time.   This was first noticed by KLEBAN[7- 8]. Unfortunately the insight did not immediately guide its originator or others to a better theory. Thus experimenters who took this "excluded volume constraint" seriously were to some extent left without a theory with which to compare their experiments.

The subject remained in this somewhat unhappy state for nearly ten years until the experiments of OHBAYASHI and his group began to be   published around 1984 [9- 10]. (Actually the experiments had been going on for a substantial period before that[11].) The new experiments provide data which can clarify the nature of the coupling to the light, the role of the excluded volume constraint and the accuracy of theories of quasiparticle dynamics. In the next section I review some general features which a theory of the light scattering process must possess including the constraint mentioned above. The third section reviews the new experiments and discusses them in the context of the general theoretical remarks of section 2. Some conclusions and discussion appear in the last section.

Springer Series in Solid-State Sciences, Vol. 79   **Elementary Excitations in Quantum Fluids**
Editors: K. Ohbayashi · M. Watabe   © Springer-Verlag Berlin, Heidelberg 1989

2. Fundamental Theoretical Considerations

We describe the interaction of the electromagnetic field with the fluid by means of the Hamiltonian H_{eff}:

$$H_{eff} = \int d\vec{r} \int d\vec{r'} \rho(\vec{r}) \vec{E}(\vec{r}) \cdot \alpha(\vec{r}, \vec{r'}) \cdot \vec{E}(\vec{r'}) \rho(\vec{r'}) \ .$$

Here $\rho(\vec{r})$ is the operator

$$\rho(\vec{r}) = \sum_i \delta(\vec{r} - \vec{r}_i)$$

where $\vec{r}_i$ is the position of the ith helium atom. $\vec{E}(\vec{r})$ is the microscopic electric field. $\alpha(\vec{r}, \vec{r'})$ is a polarizability tensor giving the local dipole moment of the medium at the point $\vec{r}$, given that there is a field at the point $\vec{r'}$. α is defined so that the contribution of the local polarizability associated with the polarizability of an isolated atom has been subtracted from it. α does not include any contributions to the polarizability which depend on the positions of more than two atoms. Such contributions undoubtedly exist but we will neglect their effects here.

From this coupling one easily finds using familiar manipulations of expressions from time dependent perturbation theory that the extinction coefficient $h(\omega)$ is given up to constants which will not interest us here by

$$h(\omega) \propto \int <H_{eff}(0) H_{eff}(t)> e^{-i\omega t} dt$$

$$\propto \int d\vec{r}_1 \int d\vec{r}_2 \int d\vec{r}_3 \int d\vec{r}_4 \hat{\epsilon}_0 \cdot \alpha(\vec{r}_1, \vec{r}_2) \cdot \hat{\epsilon}_n \hat{\epsilon}_0 \cdot \alpha(\vec{r}_3, \vec{r}_4) \cdot \hat{\epsilon}_n <\rho(\vec{r}_1, 0)\rho(\vec{r}_2, 0)\rho(\vec{r}_3, t)\rho(\vec{r}_4, t)> e^{-i\omega t} dt.$$

Here $\hat{\epsilon}_0$ and $\hat{\epsilon}_n$ are respectively the polarizations of the incoming and outgoing light in the scattering process. The problem of calculating the scattering thus separates into the problem of knowing the coupling $\alpha(\vec{r}, \vec{r'})$ and the problem of computing the four point density correlation function $< \rho(\vec{r}_1, 0)\rho(\vec{r}_2, 0)\rho(\vec{r}_3, t)\rho(\vec{r}_4, t) >$. Both aspects are more difficult than in the otherwise somewhat analogous problem of calculating the inelastic neutron scattering cross-section: In the neutron case the coupling is completely known and only a two point density correlation function is required.

With regard to the coupling, one can show [3] that the most general possible form of α in an isotropic fluid is

$$\alpha(\vec{r}, \vec{r'}) = \alpha_d \Big[\frac{3(\vec{r} - \vec{r'})(\vec{r} - \vec{r'})}{|\vec{r} - \vec{r'}|^2} - 1 \Big] + \alpha_s 1 \ .$$

The scalars α_d and α_s are functions only of the magnitude $|\vec{r} - \vec{r'}|$. Physically, one thinks of the term proportional to α_d as arising from the "dipole-induced-dipole " mechanism by which a dipole induced at position $\vec{r}$ by the incoming field produces a field which polarises the medium at the position $\vec{r'}$ through the classical dipole-dipole interaction. The dependence of α_d on $|\vec{r} - \vec{r'}|$ arises because when the atoms in question are close together their wave functions overlap and the coefficient of the classical dipole-dipole interaction becomes modified, though the tensor character of the contribution to α remains the same. The second term, proportional to α_s in α arises entirely from these non classical wave-function overlap effects. At the quantum level, these contributions to α can both be regarded as arising from a third order perturbation theory description of the interaction of the light with the electrons and nuclei which are the constituents of the fluid as illustrated schematically in Fig. 1a.

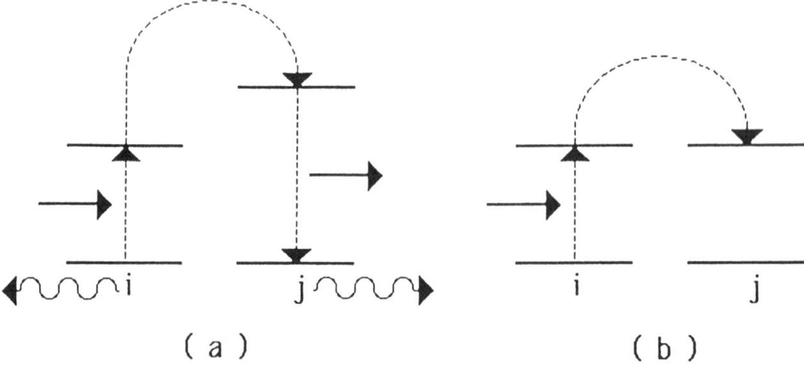

Figure 1. a) Microscopic mechanism of second order light scattering. Solid arrows denote photons, horizontal lines denote electronic energy levels associated with helium atoms at the sites i and j. Dashed arrows indicate transitions between electronic states and wavy lines show nuclear motion produced in the final state. b) Corresponding diagram describing the broadening mechanism occurring in absorption of ultraviolet light by liquid helium in the frequency region ($\approx 20\ ev$) exciting the first excited p-level in the helium atom.

In this description, the second order polarizability $\alpha(\vec{r}, \vec{r}\prime)$ arises from a sum of contributions from virtual excitation of electronic states with p-like symmetry which are transferred from one atomic site to another, followed by deexcitation with emission of the scattered photon. As described below, this point of view was used in reference 12 in order to estimate the magnitude of the scattering terms arising from the term α_s by use of data on absorption of ultraviolet light by liquid helium four.

By Fourier transform in space, one can rewrite the expressions for the extinction coefficient in the form

$$\int d^3q \int d^3q\prime t(\vec{q})t(\vec{q}\prime)G(\vec{q}, \vec{q}\prime, \omega) \, ,$$

where we have used the fact that α depends only on the vector difference between its arguments and defined

$$t(\vec{q}) = \int d^3r e^{i\vec{q}\cdot\vec{r}} \hat{\epsilon}_o \cdot \alpha(\vec{r}) \cdot \hat{\epsilon}_n \, .$$

Up to constants, $G(\vec{q}, \vec{q}\prime, \omega)$ is just the Fourier transform in space of the 4 point density correlation function of the earlier expressions with respect to the variables $\vec{r}_1 - \vec{r}_2$ and $\vec{r}_3 - \vec{r}_4$. One can write this expression in another way using the fact that the fluid is isotropic so that $G(\vec{q}, \vec{q}\prime, \omega)$ depends only on the magnitudes of $\vec{q}$ and $\vec{q}\prime$ and on the angle between them so that it may be expanded in Legendre polynomials. Further, it is not hard to show that

$$t(\vec{q}) = t_d(q)\hat{\epsilon}_o \cdot (3\hat{q}\hat{q} - 1) \cdot \hat{\epsilon}_n + t_s(q)\hat{\epsilon}_o \cdot \epsilon_n$$

in which $t_d(q) = 4\pi \int r^2 j_2(qr)\alpha_d(r)dr$ and $t_s(r) = 4\pi \int r^2 j_0(qr)\alpha_s(r)dr$. With these developments it then follows that the extinction coefficient can be written in the form

$$h(\omega) \propto I_s(\omega)(\epsilon_o \cdot \epsilon_n)^2 + I_d(\omega)[3/4 + 1/4(\hat{\epsilon}_o \cdot \hat{\epsilon}_n)^2].$$

108

Here

$$I_d(\omega) = \int 4\pi q^2 \, dq \int 4\pi q'^2 \, dq' t_d(q) t_d(q')(4/25) G_d(q, q', \omega)$$

and

$$I_s(\omega) = \int 4\pi q^2 \, dq \int 4\pi q'^2 \, dq' t_s(q) t_s(q') G_s(q, q', \omega) \, .$$

The functions $G_{s,d}(q, q', \omega)$ are the l=0,2 Legendre projections of the four point correlation function $G(\vec{q}, \vec{q'}, \omega)$:

$$G_{s,d}(q, q', \omega) = (2l + 1)/2 \int P_l(cos\gamma_{\hat{q},\hat{q'}}) G(\vec{q}, \vec{q'}, \omega) dcos\gamma_{\hat{q},\hat{q'}}, \quad l = 0, 2 \, .$$

These s and d wave components of the light scattering are a particularly useful probe of our understanding of the mechanisms of the scattering as we will discuss next.

Early theoretical work on the light scattering mechanism assumed that only the dipole-induced-dipole process described above contributed to the scattering, and therefore predicted that $I_s(\omega) = 0$ everywhere. Indeed, the experimental data for $I(\omega)$ near the two roton frequency was consistent with this prediction. In reference 12 we subjected this question to closer scrutiny by using data on the absorption of ultraviolet light in the frequency range associated with the first excited p level of the helium atom to estimate the relevant matrix elements . The idea of the calculation can be understood by reference to Fig. 1. The matrix element associated with the broadening of the ultraviolet absorption line is the same as the matrix element associated with transfer of the virtual p excitation from one site to another in the light scattering process. In making the identification, one makes a number of approximations: The light scattering amplitude is a sum over amplitudes for processes of the type illustrated in Fig. 1a in which all possible excited states of p-like symmetry are included, whereas in making the estimate from the ultraviolet width, we only include one of these terms, namely the one involving the lowest lying excited p-level. We estimated the ratio $\int I_s(\omega) d\omega / \int I_d(\omega) d\omega$. By integrating over frequency, the frequency dependent four point density correlation function in the amplitude reduced to an equal time correlation function which is somewhat easier to estimate than the frequency dependent one. Nevertheless, errors were associated with our very approximate estimates of the equal time correlation function. Finally, the estimates of the ultraviolet line width from experiment were quite uncertain. Despite these drawbacks, the method gave a definite conclusion: The observed ultraviolet line width was much too large to be accounted for by the dipole-induced-dipole interaction and, as a consequence, there had to be a substantial, observable s-wave component to the Raman scattering spectrum for some range of frequencies ω. Because no s-wave component was observed near the two roton frequency, the s-wave part of the scattering had to lie at higher frequencies. This prediction was published in 1975. Our estimate for the ratio was

$$0.11 \leq \frac{\int I_s(\omega) d\omega}{\int I_d(\omega) d\omega} \leq 1.45 \, .$$

In later work [13] the four point correlation function was estimated by more rigorous methods and two different expressions for α suggested in the literature were used. The expressions for α differed by so much that the resulting prediction for the polarization ratio covered almost the same range as that of reference 12. There is a need for better quantum chemical calculations of the functions α_d and α_s. Recent experimental results on the polarization ratio are compared with experiment in the next section and elsewhere in this volume.

With regard to the calculation of the four point time-dependent density correlation function, we first describe the excluded volume constraint [7-8] mentioned earlier. The basic constraint is very simple. It states that

$$< \rho(\vec{r}_1, 0)\rho(\vec{r}_2, 0)\rho(\vec{r}_3, t)\rho(\vec{r}_4, t) >= 0$$

when

$$|\vec{r}_1 - \vec{r}_2| < \sigma$$

or

$$|\vec{r}_3 - \vec{r}_4| < \sigma$$

in which σ is the hard core radius of a helium atom. This requirement is completely non-controversial and irrefutable. It simply requires that two helium atoms cannot be in the same place at the same time. Because it refers to positions of atoms at equal times, the problems which arise with the existing approximations with regard to this constraint are not confined to any particular frequency regime. Note that no excluded volume constraint of this type arises in analysis of neutron scattering experiments because the two factors of the density which occur in the correlation function relevant to that experiment refer to different times. In references 7- 8 it was found convenient to express the constraint in terms of a half Fourier transform:

$$C(\vec{q}, \vec{r}, \omega) = \int e^{i\vec{q}\prime \cdot \vec{r}} G(\vec{q}, \vec{q}\prime, \omega) d^3 q\prime$$

in terms of which the constraint is

$$C(\vec{q}, \vec{r}, \omega) = 0, \quad for \quad |\vec{r}| < \sigma$$

for all frequencies ω.

With this constraint in mind we now briefly review the existing techniques for calculation of the four point correlation function. The first and simplest approximation is to factor the four point correlation function:

$$< \rho(\vec{r}_1, 0)\rho(\vec{r}_2, 0)\rho(\vec{r}_3, t)\rho(\vec{r}_4, t) > \approx < \rho(\vec{r}_1, 0)\rho(\vec{r}_3, t) >< \rho(\vec{r}_2, 0)\rho(\vec{r}_4, t) > +3 \leftrightarrow 4 \quad .$$

In a sense which is made explicit in the field theoretic models to be discussed next, this approximation corresponds to assuming that the two "quasiparticle " excitations excited by the light propagate independently without interaction. It is obvious that the approximation badly violates the excluded volume condition because neither of the factors in the two terms in the approximate expression for the correlation function can possibly account for correlations between the values of the spatial arguments at equal times. On the other hand, in the first theory which used such an approximation [14], account was taken of the additional correlations in an approximate way by setting $\alpha(\vec{r}_1 - \vec{r}_2)$ and $\alpha(\vec{r}_3 - \vec{r}_4)$ equal to zero when their arguments had magnitudes less than the hard core radius of the helium atom. At the next level of sophistication, a phenomenological model for the quasiparticle excitations of the superfluid was written down which is given by the Hamiltonian

$$H = \sum_{\vec{q}} \epsilon_{\vec{q}} \alpha_{\vec{q}}^{\dagger} \alpha_{\vec{q}} + \sum_{\vec{q}, \vec{k}} g_3(\vec{q}, \vec{k}) \alpha_{\vec{k}}^{\dagger} \alpha_{\vec{q}}^{\dagger} \alpha_{\vec{k}+\vec{q}} + h.c.$$

$$+ \sum_{\vec{q}, \vec{k}, \vec{l}} g_4(\vec{q}, \vec{k}, \vec{l}) \alpha_{\vec{q}}^{\dagger} \alpha_{\vec{k}}^{\dagger} \alpha_{\vec{l}} \alpha_{\vec{q}+\vec{k}-\vec{l}} + \cdots .$$

110

The operators $\alpha_{\vec{q}}$ are boson destruction operators. There are various formal ways to derive such a quasiparticle   description of the fluid from the microscopic Hamiltonian [15- 17]. How it is done will affect the extent to which perturbation theory treating the terms in H which are of higher than quadratic order in the creation and annihilation operators will converge. To date, no serious attempt has been made in the calculations of the liquid helium Raman spectrum to relate the couplings g_3 and g_4 to microscopically derived values. The most well known calculation of this sort [4] also assumes that the connection between the quasiparticle operators $\alpha_{\vec{q}}$ and the spatial Fourier transform $\rho_{\vec{q}}$ is linear:

$$\rho_{\vec{q}} = \sqrt{Z(\vec{q})}(\alpha_{-\vec{q}} + \alpha_{\vec{q}}^{\dagger}) \ .$$

With this assumption, $Z(\vec{q})$ must be the static structure factor $S(\vec{q})$ measured by x-ray and neutron scattering experiments. In that calculation, a random phase approximation was used to calculate the light scattering spectrum under the assumption that the coupling $g_4(\vec{q}, \vec{k}, \vec{l})$ is independent of momenta and negative. This calculation predicted that a sharp peak in the scattering would be observed just below the two roton energy, as a consequence of the existence of a bound state of two rotons arising from the attractive coupling g_4. IWAMOTO [3] predicted the same qualitative result on more general, kinematical grounds. KLEBAN and HASTINGS[8] explored whether the excluded volume constraint was satisfied by the RPA result of reference 4. They found by a short calculation that the result could only satisfy the constraint if

$$Im(B_l/(1 - X_l)) = 0 \quad for \ r < \sigma$$

where

$$B_l = \sum_{\vec{q}} j_l(qr) S(q) G_2^o(q, \omega - i\delta)$$

and

$$X_l = g_4^l \sum_{\vec{q}} G_2^o(q, \omega).$$

Here $G_2^o(q, \omega - i\delta) = 1/(\omega - i\delta - 2\epsilon_q)$ is the propagator for two noninteracting quasiparticles and $\delta = 0^+$. The quasiparticle interaction g_4 has been assumed here for simplicity to depend on the angle between the momenta of the incoming and outgoing pairs of quasiparticles in the scattering event but not on the magnitudes of their momenta and

$$g_4(\vec{q}, -\vec{q}, \vec{l}) = \sum_{l,m} 4\pi g_4^l Y_l^m(\hat{q})^* Y_l^m(\hat{l})$$

defines g_4^l. A more general case is considered in reference 8. The condition must be fulfilled for all even integers l but is relevant to the case of s and d wave scattering respectively when l=0,2. It is not hard to show that the condition reduces to a fairly good approximation to

$$\sum_i \frac{j_l(q_i r) S(q_i) q_i^2}{|d\omega_q/dq|} = 0, \ \ r < \sigma \ ,$$

where the sum is over values of q which solve the equation $\omega = 2\omega_q$. One sees from this that, if $q_i\sigma$ were $<< 1$ then the calculation of the s-wave scattering amplitude would badly violate the condition while that for the d-wave scattering would not, in accordance with the intuition that, in d-wave scattering the particles "stay away from each other". Unfortunately, it is not true that $q_i\sigma << 1$ in the case of interest. $q_i \approx 1A^{-1}$ and $\sigma \approx 1A$ so the spherical Bessel function j_2 is not small and the condition is badly violated for both s and d wave scattering. Whether this defect arises from the model, the RPA approximations in the method of calculation or from the assumption that the density is linear in the quasiparticle operators is not known. The recent work of reference 17 makes a calculation possible which partially corrects these defects. We describe that work next.

In the language of the present formulation, reference 17 assumes that the (up to orthogonalization corrections which we will not discuss here)

$$\rho_B(\vec{k}) = \sqrt{Z}(\vec{k})(\alpha_{\vec{k}}^\dagger + \alpha_{-\vec{k}})$$

where the operators $\alpha_{\vec{k}}$ enter a quasiparticle Hamiltonian of the sort written earlier. The "density" $\rho_B(\vec{k})$ is related to the spatial Fourier transfrom of the particle density through the relation

$$\rho_B(\vec{k}) = \sum_i^N e^{i\vec{k}\cdot\vec{r}_i}\left(1 + \sum_{j\neq i} \vec{k}\cdot\vec{r}_{ij}\eta(r_{ij})\right).$$

Without the term involving $\eta(r)$, this would just be the spatial Fourier transform of the particle density. The added term is intended to take account of "backflow" in the variational wavefunction describing the quasiparticle excitations in a way somewhat like the one used in the original variational calculations of Feynman and Cohen. The function $\eta(r)$ was determined variationally in an approximate way in references 17. For this purpose the quasiparticle excited state was assumed to be equal to $\rho_B(\vec{k})$ times a Jastrow-like ground state and the energy was minimized with respect to parameters in the function $\eta(r)$. In fact, in order to correctly describe the observed quasiparticle spectrum, it was necessary to assume that these quasiparticles interact according to the quasiparticle Hamiltonian written down above. In the approach of references 17, however, it is possible to calculate the couplings g_3 and g_4 microscopically and they report results for g_3. A full calculation along these lines takes explicit account of the hard core of the helium atoms and should predict the light scattering spectrum correctly. We return to this in section 3.

3. Selective Review of the New Experiments

The new experimental results of the Hiroshima group can be briefly described as including these features: i) The predicted s-wave component in the scattering has been observed [9] at high frequencies and its spectral characteristics have been determined. ii) New structure has been observed [10] in the spectrum at high frequencies, well above twice the roton frequency. The fine structure is experimentally reproducible and may be associated with three and four quasiparticle processes. iii) The scattering has been observed under pressure [18] at high resolution near the two roton frequency . The most interesting new feature under high pressure is that the sharp peak observed at vapor pressure to lie below the frequency associated with two noninteracting rotons moves at high pressure to a peak lying above the frequency of two free rotons at the same pressure. This may suggest difficulties with the qualitative picture of the peak as arising from a bound state of two rotons. We discuss possible theoretical implications of each of these new results in turn.

At vapor pressure, the experimental result for the polarization ratio is

$$\frac{\int I_s(\omega)d\omega}{\int I_d(\omega)d\omega} \approx 0.10 \quad .$$

(More details concerning the polarization experiments appear elsewhere in this volume.) This result is consistent with the lower end of the range of theoretical estimates made 10 years earlier. The experiments also provide more detail on the spectral form of the s and d wave components of the scattering at high frequencies: Both components fall off approximately exponentially in the frequency with the s-wave scattering falling off more slowly with frequency than the d-wave scattering. I suggest that a convenient way to analyse the spectral shape is in terms of its frequency moments. The zeroth moment is the integrated intensity which was estimated in the earlier theoretical work. The first frequency moment of the spectrum can be calculated using a straightforward extension of procedures familiar in the derivation of the f-sum rule. One finds

$$\int h_{d,s}(\omega)\omega\,d\omega \propto 1/\pi \int [\hbar q^4/2m]t^2{}_{d,s}(q)\,N\,S(q)\,dq$$

$$+\frac{(2l+1)}{2}(4\pi)^2 \int q^3\,dq \int q\prime^3\,dq\prime t_{d,s}(q)t_{d,s}(q\prime) \int P_l(\hat{q}\cdot\hat{q}\prime)d(\hat{q}\cdot\hat{q}\prime)\frac{\hbar\hat{q}\cdot\hat{q}\prime}{m} < \rho_{\vec{q}}\rho_{-\vec{q}\prime}\rho_{\vec{q}\prime-\vec{q}} >, \quad l=0,2$$

where $S(q)$ is the structure factor. The first moment is thus a rather direct measure of the coupling to the light, involving only the coupling factors $t_{s,d}(q)$ and two static structure functions characterizing the ground state of the fluid. One of the structure functions is the well known structure factor while the other, three density correlation function, has been calculated in various approximations before. Further, one can understand qualitatively from this expression that the first moment should be larger for s than for d wave scattering, because the integrals on q giving the moment heavily weight large q where the s-wave coupling factor $t_s(q)$ is expected to be larger (because the s-wave coupling is of shorter range). Detailed comparisons of this result for the first moment with experiment have not yet been made but they are not difficult to make and results are expected soon. Higher moments are somewhat more involved [19] but a few more moments should not be difficult to calculate in order to obtain a more detailed picture of the coupling from experiment.

The experiments showing fine structure in the spectrum of the light scattering at frequencies well above the frequency of two free rotons are reviewed eleswhere in this volume. This structure, though a small effect, appears to be reproducible and there is some indication that the various features may correspond to excitation of three rotons, two maxons, two rotons and a maxon, four rotons, two maxons and a roton and three maxons. The largest feature is near the frequency of four rotons (as predicted in a calculation [20] using a variant of the formalism of reference 4 more than ten years ago. Note that this calculation almost certainly violates the excluded volume constraint.) Recently, HIRASHIMA and IWAMOTO [21] have discussed the kinematics of the proposed three quasiparticle processes. Here we discuss two possible mechanisms of three and higher quasiparticle processes in the light scattering spectrum. It is of some importance to distinguish between these. I will refer to the two mechanisms as 1) quasiparticle anharmonicity and 2) nonlinearities in the density-quasiparticle relation. In each case, I will briefly illustrate how a calculation of the fine structure arising from these mechanisms would go using the formalism of reference 17. Such calculations are under way [22], but numerical results are not yet available.

The terms involving quasiparticle interactions in the quasiparticle Hamiltonian above (involving $y_3, y_4, ...$) are what is meant by quasiparticle anharmonicities. They lead to multiquasiparticle processes in the light scattering spectrum through splitting of one or both members of the quasiparticle pair initially excited by the light. This is shown in Fig. 2a, which may be regarded as schematic or as a diagram denoting a term in time dependent perturbation theory for the rate. In terms of Feynman graphs for the response function giving the light scattering, multiparticle spectral features due to quasiparticle anharmonicity arise from series of the sort shown in Fig. 2b.

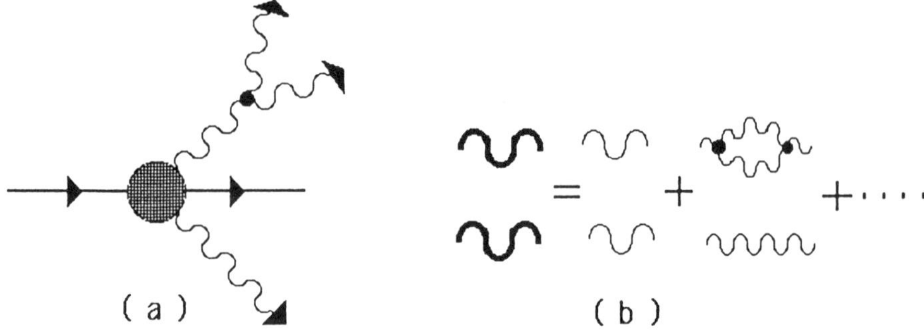

Figure 2. a) Process leading to multiquasiparticle signature arising from quasiparticle anharmonicity. A three quasiparticle process is shown. Solid arrows denote photons. Wavy arrows denote quasiparticles. b) Corresponding Feynman diagrams for the response function. Here heavy wavy lines are full quasiparticle propagators and light wavy lines are free quasiparticle propagators.

This sort of calculation led to the prediction of a four roton peak in reference 20. (Three quasiparticle signatures were predicted to be small.) In a modern version of such a calculation, one should use a formalism which makes a serious effort to take account of the excluded volume constraint, such as the formalism of reference 17.

There is another,less widely recognised source of multiquasiparticle processes in a correctly formulated quasiparticle theory, which is the nonlinearities in the relation between the density and the quasiparticle operators. We illustrate within the formalism of reference 17. Using expressions for the quasiparticle density ρ_B written in section 2 above, one easily shows that

$$\rho(\vec{k}) = \rho_B(\vec{k}) - \sum_{i \neq j} e^{i\vec{k}\cdot\vec{r}_{ij}} i\vec{k}\cdot\vec{r}_{ij}\eta(\vec{r}_{ij}) \ .$$

The last term can be written in turn in terms of two factors of the density ρ. Finally, iterating the preceding equation by substituting ρ_B for ρ in the second term one obtains

$$\rho(\vec{k}) = \rho_B(\vec{k}) + \sum_{\vec{q}} \rho_B(\vec{q})\rho_B(\vec{k}-\vec{q})M(\vec{q},\vec{k}) + ...$$

in which

$$M(\vec{q},\vec{k}) = -i \int \frac{d^3r}{V}(\vec{k}\cdot\vec{r})\eta(\vec{r})e^{i\vec{q}\cdot\vec{r}}.$$

This, together with the relation $\rho_B(\vec{k}) \propto (\alpha_{\vec{k}}^{\dagger} + \alpha_{-\vec{k}})$ generates terms describing three and more quasiparticle excitations when it is used in the basic expression for the extinction coefficient written down in section 2. This mechanism for multiquasiparticle excitation does not require that the quasiparticles interact at all. As discussed in reference 8 , these nonlinear terms in the density-quasiparticle relation are likely to be required in order to satisfy the excluded volume constraint. In the present formulation, based on the formalism of reference 17 one sees the connection to the excluded volume constraint in a very physical way, because the terms involving η in the relation of ρ_B to ρ are thought physically to arise to account for "backflow" which occurs in order that the particles not overlap during the motion associated

114

with the excitation. For these reasons and because the simple interacting quasiparticle models without these corrections badly violate the excluded volume constraint, we expect the contributions of these nonlinear terms to the multiparticle scattering rate to be large. Only detailed calculation will show whether they are larger than the multiquasiparticle amplitudes arising from quasiparticle anharmonicity.

Finally we briefly address the new high resolution experiments in the neighborhood of twice the roton frequency at high pressure. At these frequencies, the corrections arising from the multiparticle terms arising from the nonlinearities in the quasiparticle density relation are not expected to give rise to sharp structure but could lead to a background term which was slowly varying in the frequency. Such corrections might account for the (rather small) discrepancies between the observed spectral shape and phenomenological fits to the theory of reference 4. A second possible origin of these discrepancies is the momentum dependence of g_4. We showed in reference 20 that such momentum dependence could lead to modifications of the shape of the "bound state" peak somewhat like what is observed experimentally. To check whether this is the correct origin of the discrepancy requires a microscopic calculation of g_4. Finally, it is possible that part of the discrepancy arises because of the inadequacy of the ladder approximation. Leading corrections to vertex function in the Bethe-Saltpeter equation would not be too difficult to   evaluate in order to explore this possibility [23].

4. Conclusions

We have reviewed the theory and the new experiments in the light of the existing experimental situation. We conclude that a closer theoretical analysis of the existing new experiments can yield much more detailed information about the coupling than has heretofore been available. To be most valuable, such a program needs to include new quantum chemical calculations of the function α as well as calculations of several moments of the response function. The high frequency fine structure can yield information about the nature of the nonlinear terms in the quasiparticle-density relation and in particular can provide a test of the form for these terms proposed in reference 17. Finally, the new high resolution studies at high pressure are most likely to yield information about the momentum dependence of the quasiparticle coupling g_4 which can be compared with theories such as that of reference 17.

5. References

1.   T. Greytak and J. Yan, <u>Phys. Rev. Lett.</u>22, 987 (1969)

2.   J. W. Halley, <u>Bull. Am. Phys. Soc. 13</u>, 398 (1968);
     <u>Phys. Rev.181</u>, 338 (1969)

3.   F. Iwamoto, <u>Prog. Theo. Phys.44</u>, 1135 (1970)

4.   J. Ruvalds and A. Zawadowski, <u>Phys. Rev. Lett. 25</u>, 333 (1970);
     A. Zawadowski, J. Ruvalds, and J. Solana, <u>Phys. Rev.</u> A5399 (1972)

5.   T. J. Greytak, R. L. Woerner, J. Yan and R. Benjamin,
     <u>Phys. Rev. Lett. 25</u>, 1547 (1970)

6.   C. A. Murray, R. L. Woerner, and T. J. Greytak, <u>J. Phys. C 8</u>, L90 (1975)

7.   P. Kleban, <u>Phys. Lett.</u> <u>49A</u>, 19 (1974); <u>Phys. Rev.</u> <u>B19</u>, 3511 (1979)

8.   P. Kleban and R. Hastings, <u>Phys. Rev.</u> <u>B11</u>, 1878 (1975)

9.   M. Udagawa, H. Nakamura, M. Murakami and K. Ohbayashi,
      <u>Phys. Rev.</u> <u>B 34</u>, 1563 (1986)

10.   K. Ohbayashi and M. Udagawa, <u>Phys. Rev.</u> <u>B31</u>,1324(1985)

11.   K. Ohbayashi and A .Ikushima, <u>J. Phys.</u> <u>C 7</u>,L206 (1974)

12.   P. Kleban and J.W. Halley, <u>Phys. Rev.</u> <u>B11</u>, 3520 (1975)

13.   C. E. Campbell, J. W. Halley and F. Pinski, <u>Phys. Rev.</u> <u>B21</u>,1323 (1980)

14.   M. J. Stephen, <u>Phys. Rev.</u> <u>187</u>,279 (1969)

15.   H. W. Jackson, <u>Phys. Rev.A8</u>,1529 (1973)

16.   S. Sunakawa, S. Yamasaki and T. Kebukawa,
      <u>Prog. Theor. Phys.</u> <u>41</u>,919 (1969)

17.   E.Manousakis and V. Pandharipande,
      <u>Phys. Rev.</u> <u>B31</u>,7029(1985); <u>B33</u>,150(1986)

18.   K. Ohbayashi, T. Akagi, N. Ogita M. Watabe and M. Udagawa,
      Proceedings of LT18 (in press)

19.   J. W. Halley and C. Campbell (unpublished)

20.   R. Hastings and J. W. Halley, <u>Phys. Rev.</u> <u>A 10</u>,2488 (1974)

21.   D. S. Hirashima and F. Iwamoto, <u>Prog. Theor, Phys.</u> <u>75</u>,744 (1986),
      see also the paper by Iwamoto in this volume.

22.   J. W. Halley and M. Korth (unpublished)

23.   J. W. Halley in <u>Correlation Functions and Quasiparticle Interactions
      in Condensed Matter</u>, J. W. Halley, ed.,Plenum, NY (1978), p. 87

Effect of the Three-Excitation States on the Raman Scattering from Superfluid ^{4}He

Fumiaki Iwamoto

Institute of Physics, College of Arts and Sciences, University of Tokyo, Komaba, Tokyo 153, Japan

1. Introduction

Several years ago, OHBAYASHI and UDAGAWA [1] told us that they had seen in the Raman spectrum an abrupt change of the slope at the points a, b, c, d, e, and f as well as peaks at p_1 and p_2.

Frankly speaking, I did not believe these structures to be real except those at p_1, d, and p_2, which nearly correspond to the energy transfer $2\Delta_0$, $2\Delta_1$, and $2\Delta_2$. Here, Δ_0, Δ_1, and Δ_2 denote the energies of the minimum, the maximum, and the plateau of the phonon–roton dispersion curve. Then, I suggested to them that they should work with sharper resolution at a lower temperature. But, Ohbayashi strongly insisted on the structure: He said, "Even with this resolution and even at this temperature I can see some structure. The real structure must be very sharp." And he suggested to me that I should reexamine the theory.

So, I reexamined the theory with HIRASHIMA [2]. It was hard for us to derive some irregularities from the continuous dispersion curve. The only source, which we could think of, was the effects from the channels of more than two excitations. The theories [3,4] so far had been restricted to the two-excitation states.

In the next section, a framework is given for the description of the elementary excitations following the spirit of LANDAU [5]. To help the physical understanding of the behavior of the Raman spectrum near $2\Delta_0$ and $2\Delta_1$, I remark in Sect.3 on some properties of the two excitations produced by light scattering. It is shown in Sect.4 that the density of three-excitation states has a term which behaves like $\sqrt{E - E_c}$ at the energy E above $E_c = 3\Delta_0$ and $\Delta_0 + 2\Delta_1$, and $\sqrt{E_c - E}$ below $E_c = 2\Delta_0 + \Delta_1$ and $3\Delta_1$. Just like the WIGNER [6] cusp, it is expected that this irregular density of states naturally leads to the cusp or the inflection in the Raman spectrum. To confirm the expectation, I calculate in Sect.5 the Raman matrix elements taking into account the interaction among the two- and three-excitation states. Finally, by giving a formula for the Raman spectrum in Sect.6, I conclude the existence of the cusp or the inflection with infinite derivative in the Raman spectrum just at $E = 3\Delta_0$, $2\Delta_0 + \Delta_1$, $\Delta_0 + 2\Delta_1$ and $3\Delta_1$ as well as at $2\Delta_0$ and $2\Delta_1$.

Throughout this report I use units where $\hbar = 1$ and (the volume of the system) $= 1$.

2. Description of Elementary Excitations

Thanks to Landau we can, to a certain extent, treat the superfluid ^{4}He as a dilute gas of elementary excitations rather than as a complicated many-body system of helium atoms. The neutron scattering experiment determines the dispersion curve $\omega(p)$, which is the energy-momentum relation of an elementary excitation. As was studied by PITAEVSKII [7], this curve has a remarkable property: When we disregard a phonon of long wavelength and an excitation near the end of the plateau, an excitation cannot decay into many excitations from the conservation of energy and momentum. This fact means that the one-excitation state is an exact eigenstate of the hamiltonian of the system. I write $|0\rangle$ for the ground state and $b_{\vec{p}}^{\dagger}|0\rangle$ for the one-excitation state of momentum $\vec{p}$. The operators $b_{\vec{p}}^{\dagger}$ and their hermitian conjugates $b_{\vec{p}}$ can be required to satisfy the boson commutation relations. It is to be noted that such a set of operators certainly exists, but its uniqueness is a question. I return to this point at the end of this report.

In general, excitations not only scatter to each other but also change their number. Therefore, a free n-excitation state simply labeled by α,

$$|\alpha\rangle = \frac{1}{\sqrt{n_1! \cdots n_\ell!}} (b_{\vec{p}_1}^{\dagger})^{n_1} \cdots (b_{\vec{p}_\ell}^{\dagger})^{n_\ell} |0\rangle \ ,$$

$$n = n_1 + \cdots + n_\ell \geq 2, \tag{2.1}$$

is not the eigenstate of the hamiltonian. The hamiltonian naturally splits into two parts, $H = H_0 + V$; the free hamiltonian H_0 and the interaction V,

$$H_0 = \sum_{\vec{p}} \omega(p) b_{\vec{p}}^{\dagger} b_{\vec{p}} \ , \tag{2.2}$$

$$V = \sum_{\alpha,\beta} |\alpha\rangle V(\alpha,\beta) \langle\beta| \ . \tag{2.3}$$

The interacting eigenstate Ψ_A is a superposition of states of two and more excitations,

$$\Psi_A = \sum_\alpha f_A(\alpha) \, |\alpha\rangle. \tag{2.4}$$

The amplitude $f_A(\alpha)$ satisfies the following equation:

$$(E - \omega_\alpha)f_A(\alpha) = \sum_\beta V(\alpha,\beta)f_A(\beta) \, , \tag{2.5}$$

where E is the energy of the state Ψ_A, and $\omega_\alpha = n_1\omega(p_1) + \cdots + n_\ell\omega(p_\ell)$ is the sum of energies of the free excitations in the state $|\alpha\rangle$.

Further pursuit of this general line will be fruitless. Approximations should be introduced according to each physical situation.

3. Two-Excitation State

HALLEY [8] was the first to study the Raman scattering of superfluid ^{4}He. A simplifying feature is that the total momentum of the produced excitations can be taken as zero, because visible light carries negligible momentum. He considered the creation of two excitations of momenta $\vec{p}$ and $-\vec{p}$, and calculated the density $\rho_2(E)$ of the two-excitation states,

$$\rho_2(E) = \frac{1}{2 \cdot (2\pi)^3} \int \delta(2\omega(p) - E)d\vec{p}$$

$$= \frac{1}{2 \cdot (2\pi)^3} \sum_\lambda \left(\frac{4\pi p^2}{2|d\omega/dp|}\right)_{p=p_\lambda} , \tag{3.1}$$

where p_λ is the solution of $\omega(p_\lambda) = E/2$. Figure 1 shows that there are three solutions p_a, p_b, p_c when $\Delta_0 < E/2 < \Delta_1$, and one solution p_a or p_c when $E/2 < \Delta_0$ or $E/2 > \Delta_1$. By taking $\omega(p)$ as parabolic near the minimum,

$$\omega(p) = \Delta_0 + \frac{1}{2\mu_0}(p - P_0)^2, \tag{3.2}$$

and near the maximum,

$$\omega(p) = \Delta_1 - \frac{1}{2\mu_1}(p - P_1)^2, \tag{3.3}$$

he found that $\rho_2(E)$ diverges as $1/\sqrt{E - 2\Delta_0}$ above $2\Delta_0$, and as $1/\sqrt{2\Delta_1 - E}$ below $2\Delta_1$. Therefore, he expected that the Raman spectrum has peaks at $E = 2\Delta_0$ and $2\Delta_1$.

The earliest experiment by GREYTAK and YAN [9] showed a large peak at about $2\Delta_0$, but no evidence of the possible peak at $2\Delta_1$. The non-existence was a puzzle.

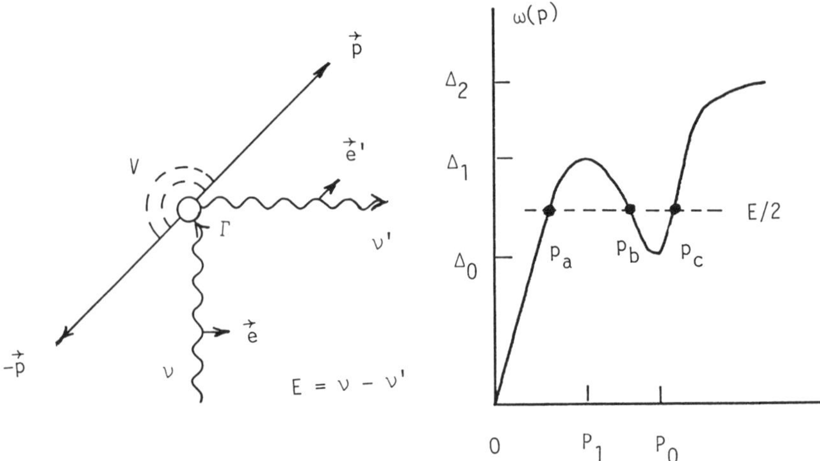

Fig.1 Energies and momenta of two excitations produced by the inelastic
light scattering

It turned out that the final state interaction between the produced excitations gives the answer to the puzzle. Because it was fully analyzed in the references [3,4] and the formula to be given later in this report contains the effect as a special case, here I make four physical remarks, all connected with the properties of the dispersion relation.

(i) When the produced excitations are near the minimum or the maximum of the dispersion curve, their group velocity $d\omega/dp$ is very small. The final state interaction is very effective for such slow excitations.

(ii) The group velocity is negative between the maximum and the minimum. The momentum of the excitation and the velocity of its wave packet point in opposite directions to each other. When $\Delta_0 < E/2 < \Delta_1$, there are three open channels; three kinds of waves p_a, p_b, p_c reach the wave zone. Close examination [10,3] of the wave function of a pair of excitations of total momentum zero shows that it behaves as

$$\psi_a(\vec{r}) \sim e^{i\vec{p}_a \cdot \vec{r}} + f_{aa}\frac{e^{ip_a r}}{r} - f_{ba}\frac{e^{-ip_b r}}{r} + f_{ca}\frac{e^{ip_c r}}{r} \qquad (3.4)$$

for the incident plane wave in the channel-a. In the channel-b, where the group velocity is negative, the outgoing spherical wave is represented by $-e^{-ipr}/r$, not by the ordinary e^{ipr}/r. A similar change of sign appears when the incident wave is in the other two channels. This change of sign causes a destructive interference between the channel-b and the channel-c when the

120

energy E is slightly larger than $2\Delta_0$, and between the channel-a and the
channel-b when E is slightly smaller than $2\Delta_1$. Due to this interference the
Raman spectrum does not show peaks at $2\Delta_0$ and $2\Delta_1$ where the density of
states diverges.

(iii) Similar to the mechanism of the formation of the Cooper pair in a
superconductor, a resonance should appear at an energy below $2\Delta_0$, however
small the attractive interaction may be between the excitations near the
minimum. The same happens above $2\Delta_1$, if the interaction is repulsive near
the maximum. It is a resonance, not a bound state; a large number of states
around the minimum (maximum) form a bound state, and this bound state decays
into the open channel-a (channel-c) conserving energy and momentum. Indeed,
this resonance was very clearly observed just below $2\Delta_0$ by MURRAY et al.
[11].

(iv) Due to the isotropy of the dispersion relation, the total angular
momentum L is a good quantum number for a many-excitation state of total
momentum zero. Because a photon has spin 1 and its orbital angular momentum
is zero for the present purpose, the conservation of the angular momentum
for the Raman process is expressed by

$$\vec{L} + \vec{1} = \vec{1}. \tag{3.5}$$

We have L = 0 or 2, because L = 1 is excluded from the Bose statistics or
from the parity conservation. A depolarization experiment [9,12,1] on the
scattered light tells us the relative amounts of the S-state and the D-state
of the produced excitations.

Incidentally, it may be of some historical interest to look back at the
origin of the name "roton". In 1941 Landau conceived of an excitation
corresponding to some kind of rotational motion. He called it a roton after
Tamm's suggestion and conjectured its energy-momentum relation,

$$\omega(p) = \Delta + \frac{p^2}{2\mu} . \tag{3.6}$$

However, he abandoned the excitation in 1947, and switched the name to the
excitations around the minimum of the dispersion curve which follows (3.2).
Therefore, the roton-'47 lost its meaning as rotation, and became a phonon
of short wavelength. It is to be noted that the resonance mentioned above
has all the properties of the roton-'41: It obeys [13] the energy-momentum
relation (3.6), and rotates in a D-state.

4. Density of Three-Excitation States

In this section I calculate the density $\rho_3(E)$ of three-excitation states of total momentum zero, and show that it contains a term which behaves as $\sqrt{|E - E_c|}$ on one side of $E_c = 3\Delta_0$, $2\Delta_0 + \Delta_1$, $\Delta_0 + 2\Delta_1$, and $3\Delta_1$.

Because the total momentum is zero, the momenta of three excitations $\vec{p}_1$, $\vec{p}_2$, and $\vec{p}_3 = -\vec{p}_1 - \vec{p}_2$ form a triangle as in Fig.2. The density of states is

$$\rho_3(E) = \frac{1}{3!(2\pi)^6} \int \delta(\omega(p_1) + \omega(p_2) + \omega(|\vec{p}_1 + \vec{p}_2|) - E)\, d\vec{p}_1 d\vec{p}_2. \qquad (4.1)$$

The factor $1/3!$ comes from the Bose statistics. I transform the integration variables to the magnitudes of three momenta p_1, p_2, p_3 and the Euler angles θ, ϕ, ψ, which specify the orientation of the triangle. Angular integration gives $8\pi^2$, and $\rho_3(E)$ becomes

$$\rho_3(E) = \frac{8\pi^2}{3!(2\pi)^6} \int \delta(\omega(p_1) + \omega(p_2) + \omega(p_3) - E)\, p_1 p_2 p_3\, dp_1 dp_2 dp_3 \qquad (4.2)$$

where $p_1 p_2 p_3$ comes from the Jacobian, and the integration region is limited by the triangular inequality,

$$|p_1 - p_2| \leq p_3 \leq p_1 + p_2. \qquad (4.3)$$

When p_1, p_2, p_3 are all near the minimum of the dispersion curve represented by (3.2), they give a contribution

$$\rho_{3\Delta_0}(E) = \frac{8\pi^2 \cdot 2\mu_0}{3!(2\pi)^6} \int \delta((p_1 - P_0)^2 + (p_2 - P_0)^2 + (p_3 - P_0)^2 - R^2)$$

$$\times\, p_1 p_2 p_3\, dp_1 dp_2 dp_3, \qquad (4.4)$$

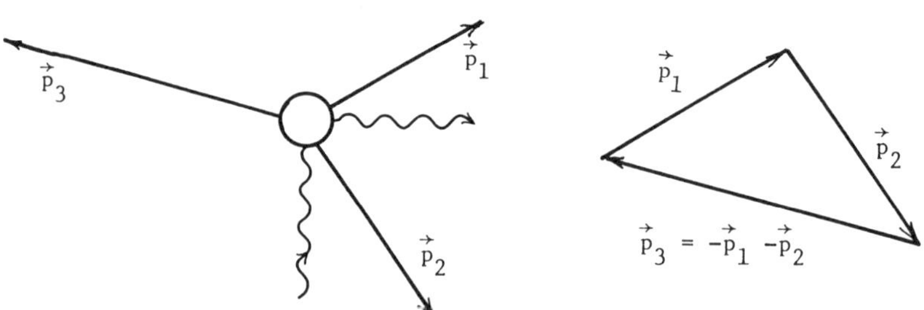

Fig.2 Triangle of the produced three excitations

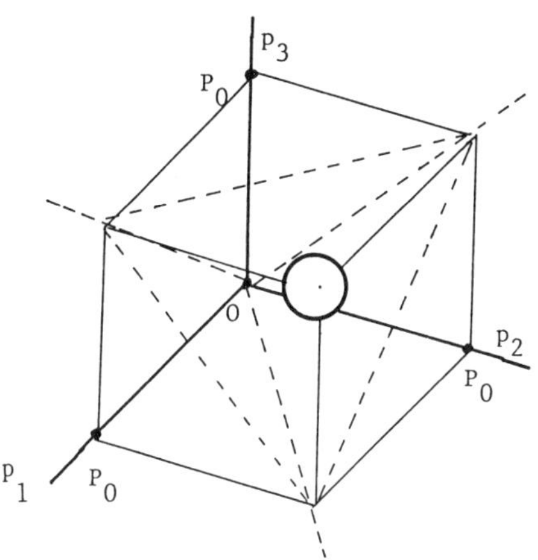

Fig.3 Equi-energy surface,
a sphere at (P_0, P_0, P_0)
in the trigonal pyramid

when $E > 3\Delta_0$. As shown in Fig.3, the equi-energy surface in the $p_1p_2p_3-$ space is a sphere of radius $R = \sqrt{2\mu_0(E - 3\Delta_0)}$ with its center at (P_0, P_0, P_0). When R is not so large, the sphere is safely in the region (4.3), which forms a trigonal pyramid shown in Fig.3. By shifting the origin to the center and using spherical coordinates, the integral is easily calculated to give

$$\rho_{3\Delta_0}(E) = \frac{8\pi^2 \cdot 2\mu_0 P_0^3}{3!(2\pi)^6} \int \delta(r^2 - R^2)\ 4\pi r^2 dr = \frac{P_0^3\mu_0}{12\pi^3}\sqrt{2\mu_0(E - 3\Delta_0)}\ .$$

$$(4.5)$$

In a similar way I calculate the contribution $\rho_{2\Delta_0+\Delta_1}(E)$ from the region where two excitations are near the minimum and one excitation is near the maximum. By inserting (3.2), (3.3) into (4.2) and noting three identical regions, it is expressed by

$$\rho_{2\Delta_0+\Delta_1}(E) = \frac{8\pi^2 \cdot 2\mu_0}{2(2\pi)^6} \int \delta((p_1 - P_0)^2 + (p_2 - P_0)^2 - \frac{\mu_0}{\mu_1}(p_3 - P_1)^2 - a)$$

$$\times\ p_1p_2p_3\ dp_1dp_2dp_3\ . \qquad (4.6)$$

where $a = 2\mu_0(E - 2\Delta_0 - \Delta_1)$. As shown in Fig.4, the equi-energy surface is a hyperboloid of revolution with its center at (P_0, P_0, P_1). I cut off $|p_3 - P_1| \leq q_c$, so that the approximate formula (3.3) is valid. When the energy E is near $2\Delta_0 + \Delta_1$, the cut off hyperboloid is certainly in the region (4.3). The integral in (4.6) becomes

$$\pi P_0^2 P_1 \int \delta(\xi - \frac{\mu_0}{\mu_1} z^2 - a) d\xi dz,$$

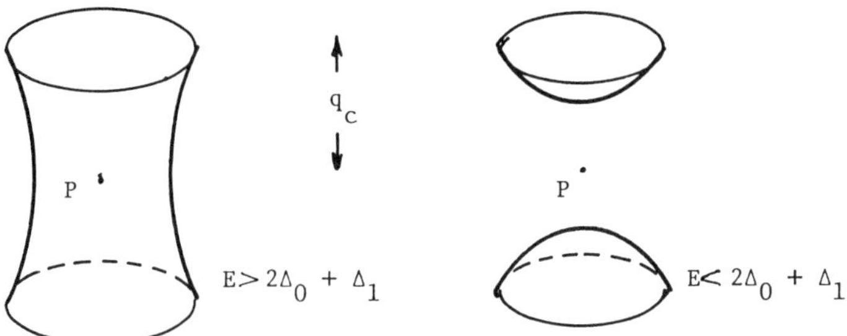

Fig.4 Equi-energy surface, a hyperboloid at $P(P_0, P_0, P_1)$

by changing the variables to $\xi = (p_1 - P_0)^2 + (p_2 - P_0)^2$ and $z = p_3 - P_1$. The ξz-integral is calculated as

$$\int \delta(\xi - \frac{\mu_0}{\mu_1}z^2 - a)d\xi dz = \int_{-q_c}^{q_c} \theta(\frac{\mu_0}{\mu_1}z^2 + a)dz$$

$$= \left\{ \begin{array}{ll} 2q_c , & a > 0 \\ 2q_c - 2\sqrt{\mu_1|a|/\mu_0} , & a < 0 . \end{array} \right.$$

Therefore, I have

$$\rho_{2\Delta_0+\Delta_1}(E) = \left\{ \begin{array}{ll} C , & E > 2\Delta_0 + \Delta_1 \\ C - \frac{P_0^2 P_1 \mu_0}{4\pi^3}\sqrt{2\mu_1(2\Delta_0+\Delta_1-E)} , & E < 2\Delta_0 + \Delta_1 . \end{array} \right. \quad (4.7)$$

The cut off momentum q_c affects only the constant $C = P_0^2 P_1 \mu_0 q_c/(4\pi^3)$.

The same calculation shows that $\rho_3(E)$ has a term $-\sqrt{E - \Delta_0 - 2\Delta_1}$ above $\Delta_0 + 2\Delta_1$, and $\sqrt{3\Delta_1 - E}$ below $3\Delta_1$. The behavior is schematically shown in Fig.5.

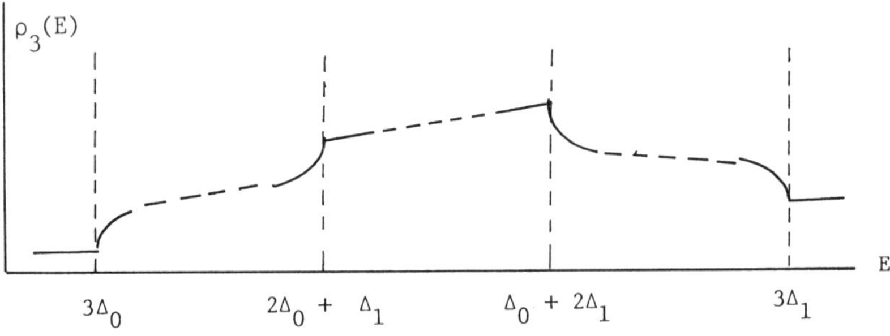

Fig.5 Schematic of $\rho_3(E)$ near $3\Delta_0$, $2\Delta_0 + \Delta_1$, $\Delta_0 + 2\Delta_1$, $3\Delta_1$

5. Raman Matrix Element

As described in Section 2, free n-excitation states ($n \geq 2$) are coupled. I approximate the interacting two- and three-excitation state Ψ_A of total momentum zero as a superposition of free two-excitation states $b_{\vec{p}}^{+} b_{-\vec{p}}^{+}|0\rangle$ and free three-excitation states $b_{\vec{q}}^{+} b_{\vec{q}'}^{+} b_{-\vec{q}-\vec{q}'}^{+}|0\rangle$, neglecting components of more than three excitations,

$$\Psi_A = \sum_{\vec{q}} f_A(\vec{q}) \, b_{\vec{q}}^{+} b_{-\vec{q}}^{+}|0\rangle + \sum_{\vec{q},\vec{q}'} g_A(\vec{q},\,\vec{q}') \, b_{\vec{q}}^{+} b_{\vec{q}'}^{+} b_{-\vec{q}-\vec{q}'}^{+}|0\rangle \, , \qquad (5.1)$$

with an obvious change of notation from Sect.2. To avoid complications in counting the identical states, I understand the sum over $\vec{q}$ or $\vec{q},\,\vec{q}'$ to be done only once for identical states. For example, I can restrict the domain to $q_z > 0$ or $q > q' > |\vec{q}+\vec{q}'|$. The approximation will be permitted, at least in order to discuss the irregular properties of the Raman spectrum at E_c, because the density of states of more than three excitations is a smooth function at E_c.

By using a certain operator Q, the Raman matrix element M_A from the ground state to the state Ψ_A is given by

$$M_A = \langle\Psi_A,\, Q\,|0\rangle = \sum_{\vec{q}} f_A^{*}(\vec{q})\Gamma_2(\vec{q}) + \sum_{\vec{q},\vec{q}'} g_A^{*}(\vec{q},\,\vec{q}')\Gamma_3(\vec{q},\,\vec{q}'). \qquad (5.2)$$

The quantities $\Gamma_2(\vec{q})$ and $\Gamma_3(\vec{q},\vec{q}')$ are the matrix elements of the free two- and three-excitation states,

$$\Gamma_2(\vec{q}) = \langle 0|b_{-\vec{q}}b_{\vec{q}}\, Q|0\rangle,$$

$$\Gamma_3(\vec{q},\,\vec{q}') = \langle 0|b_{-\vec{q}-\vec{q}'}b_{\vec{q}'}b_{\vec{q}}\, Q|0\rangle. \qquad (5.3)$$

In this way I separate the Raman matrix element into the free matrix element and the interacting amplitude.

The amplitudes satisfy the coupled equation (2.5), which is written for the present purpose with a slight change of notation,

$$\left[\begin{array}{l} \{E - 2\omega(q)\}\, f_A(\vec{q}\,) = T_2(\vec{q}\,) \\[2mm] \{E - \omega_3(\vec{q},\,\vec{q}')\}\, g_A(\vec{q},\,\vec{q}') = T_3(\vec{q},\,\vec{q}'), \end{array}\right. \qquad (5.4)$$

with

$$\left[\begin{array}{l} T_2(\vec{q}) = \sum_{\vec{s}} V_2(\vec{q};\vec{s})f_A(\vec{s}) + \sum_{\vec{s},\vec{s}'} V_{23}(\vec{q};\vec{s},\vec{s}')g_A(\vec{s},\vec{s}') \\[2mm] T_3(\vec{q}, \vec{q}') = \sum_{\vec{s}} V_{32}(\vec{q},\vec{q}';\vec{s})f_A(\vec{s}) + \sum_{\vec{s},\vec{s}'} V_3(\vec{q},\vec{q}';\vec{s},\vec{s}')g_A(\vec{s},\vec{s}'). \end{array}\right. \tag{5.5}$$

In the above, $\omega_3(\vec{q},\vec{q}') = \omega(q) + \omega(q') + \omega(|\vec{q}+\vec{q}'|)$.

Because the interacting eigenstate Ψ_A is a scattering state in the continuous spectrum, the quantum number A is specified by the momenta of the incoming excitations and its energy E becomes the sum of the energies of the free incoming excitations. Under the boundary condition of outgoing spherical waves for two incoming excitations $\vec{p}$ and $-\vec{p}$, an integral equation follows from (5.4),

$$\left[\begin{array}{l} f_{\vec{p}}^{(+)}(\vec{q}) = \delta_{\vec{q},\vec{p}} + \dfrac{1}{E - 2\omega(q) + i\varepsilon}\, T_2(\vec{q}) \\[4mm] g_{\vec{p}}^{(+)}(\vec{q}, \vec{q}') = \dfrac{1}{E - \omega_3(\vec{q},\vec{q}') + i\varepsilon}\, T_3(\vec{q}, \vec{q}'). \end{array}\right. \tag{5.6}$$

Similarly, for the three incoming excitations of $\vec{p}$, $\vec{p}'$ and $-\vec{p}-\vec{p}'$, the integral equation is

$$\left[\begin{array}{l} f_{\vec{p},\vec{p}'}^{(+)}(\vec{q}) = \dfrac{1}{E - 2\omega(q) + i\varepsilon}\, T_2(\vec{q}) \\[4mm] g_{\vec{p},\vec{p}'}^{(+)}(\vec{q},\vec{q}') = \delta_{\vec{p},\vec{q}}\delta_{\vec{p}',\vec{q}'} + \dfrac{1}{E - \omega_3(\vec{q},\vec{q}') + i\varepsilon}\, T_3(\vec{q},\vec{q}'). \end{array}\right. \tag{5.7}$$

When the quantities Γ_2, Γ_3 and V_2, V_3, $V_{23} = V_{32}^{*}$ are given, the Raman matrix element is calculated by (5.2) after solving the integral equation (5.6), (5.7). The determination of these quantities is a difficult problem. For example, the explicit expression for the operator Q is known to be

$$Q = \frac{1}{2c^2\sqrt{\nu\nu'}}\{\vec{J}\cdot\vec{e}'\,\frac{1}{\nu - H}\,\vec{J}\cdot\vec{e} - \vec{J}\cdot\vec{e}\,\frac{1}{\nu'+H}\,\vec{J}\cdot\vec{e}'\}, \tag{5.8}$$

where $\vec{e}$, ν are the polarization vector and the frequency of the incident light, and $\vec{e}'$, ν' are those of the scattered light. $\vec{J}$ is the electric current operator of the system. However, this expression is not so useful for the determination of Γ, because little is known about the relation of an elementary excitation to the helium atoms — much less the relation to the electrons. The same happens to the interaction V. The complex many-body problem prevents their determination.

126

Nevertheless, it is a reasonable assumption that these quantities are smooth functions of their arguments. Moreover, these quantities can be taken as constant as far as the irregularity of the spectrum at E_c is concerned: Expansion of the smooth function at $q = P_0$ or P_1 shows that the most irregular behavior of the spectrum comes from the constant term, and the higher order terms with respect to $q - P_0$ or $q - P_1$ give less irregular behavior.

Attention should be paid to the dependence on the orientation of the vectors $\vec{e}$, $\vec{e}'$ $\vec{q}$, $\vec{q}'$, $\vec{p}$, $\vec{p}'$. Actually, D-state production is dominant in the energy region of present interest [1,9,12], while the constant Γ and V lead only to S-state production. However, it is known from the analysis in [3,4] that both states show the same irregularity in the spectrum. Although the detailed analysis is complicated, such a property will certainly continue to hold even in the case of coupled two- and three-excitations. This property is different from the usual Wigner cusp at the inelastic threshold. The difference comes from the momenta of the produced particles. In the usual scattering case, the produced particles have small momenta and only the S-state wave function is nonvanishing at the interaction center. In the present case, the momenta of the produced excitations near E_c are of magnitude P_0 or P_1.

It is easy to solve the integral equation for the constant interaction. When V_2, V_3, $V_{23} = V_{32}^*$ are constant, T_2 and T_3 are also constant for given E as can be seen from (5.5). By inserting (5.6) into (5.5), T_2 and T_3 are determined by

$$
\left[
\begin{aligned}
&\{1 - V_2 F_2(E)\} T_2 - V_{23} F_3(E) T_3 = V_2 \\
&- V_{32} F_2(E) T_2 + \{1 - V_3 F_3(E)\} T_3 = V_{32},
\end{aligned}
\right.
\tag{5.9}
$$

where $F_2(E)$ and $F_3(E)$ represent

$$
F_2(E) = \sum_{\vec{q}} \frac{1}{E - 2\omega(q) + i\varepsilon} ,
\tag{5.10}
$$

$$
F_3(E) = \sum_{\vec{q},\vec{q}'} \frac{1}{E - \omega_3(\vec{q},\vec{q}') + i\varepsilon} .
\tag{5.11}
$$

Solving T_2 and T_3 from (5.9) and inserting them into (5.6), I have the solution of the integral equation (5.6),

$$\left[\begin{array}{l} f_{\vec{p}}^{(+)}(\vec{q}) = \delta_{\vec{q},\vec{p}} + \dfrac{V_2 + (|V_{23}|^2 - V_2 V_3)F_3(E)}{D(E)(E - 2\omega(q) + i\varepsilon)} \\[2em] g_{\vec{p}}^{(+)}(\vec{q},\vec{q}') = \dfrac{V_{32}}{D(E)(E - \omega_3(\vec{q},\vec{q}') + i\varepsilon)} \end{array}\right. , \qquad (5.12)$$

where $D(E)$ represents

$$D(E) = (1 - V_2 F_2(E))(1 - V_3 F_3(E)) - |V_{23}|^2 F_2(E)F_3(E). \qquad (5.13)$$

In the same way, I have the solution of the integral equation (5.7).

$$\left[\begin{array}{l} f_{\vec{p},\vec{p}'}^{(+)}(\vec{q}) = \dfrac{V_{23}}{D(E)(E - 2\omega(q) + i\varepsilon)} \\[2em] g_{\vec{p},\vec{p}'}^{(+)}(\vec{q},\vec{q}') = \delta_{\vec{p},\vec{q}}\,\delta_{\vec{p}',\vec{q}'} + \dfrac{V_3 + (|V_{23}|^2 - V_2 V_3)F_2(E)}{D(E)(E - \omega_3(\vec{q},\vec{q}') + i\varepsilon)} \end{array}\right. . \qquad (5.14)$$

As was shown by WATSON [14], the correct matrix element is given by $\langle\Psi_A^{(-)},\ Q\,|0\rangle$ when the final state interaction is present. $\Psi_A^{(-)}$ is the incoming spherical wave solution so that the quantum number A represents the final momenta of the produced excitations. Because $[f_A^{(-)}(\vec{q})]^* = f_A^{(+)}(\vec{q})$ and $[g_A^{(-)}(\vec{q},\vec{q}')]^* = g_A^{(+)}(\vec{q},\vec{q}')$, I have finally the Raman matrix elements for constant Γ from (5.2), (5.12) and (5.14). They are for $A = (\vec{p},\ -\vec{p})$

$$M_{\vec{p}} = \frac{1}{D(E)}\{\Gamma_2 + (\Gamma_3 V_{32} - \Gamma_2 V_3)F_3(E)\}, \qquad (5.15)$$

and for $A = (\vec{p},\ \vec{p}',\ -\vec{p}-\vec{p}')$

$$M_{\vec{p},\vec{p}'} = \frac{1}{D(E)}\{\Gamma_3 + (\Gamma_2 V_{23} - \Gamma_3 V_2)F_2(E)\}. \qquad (5.16)$$

6. Raman Spectrum

From the matrix elements (5.15, 16), the golden rule gives the transition probability dW for a photon of momentum $\vec{k}$ to be scattered to a photon of momentum $\vec{k}' \sim \vec{k}'+d\vec{k}'$,

$$\begin{aligned} dW &= 2\pi \frac{d\vec{k}'}{(2\pi)^3}\ \Big\{ \sum_{\vec{p}} |M_{\vec{p}}|^2 \delta(2\omega(p)-E) + \sum_{\vec{p},\vec{p}'} |M_{\vec{p},\vec{p}'}|^2 \delta(\omega_3(\vec{p},\vec{p}')-E) \Big\} \\[1em] &= \frac{d\vec{k}'}{(2\pi)^2}\ \frac{1}{|D(E)|^2}\ \{\,|\Gamma_2 + (\Gamma_3 V_{32} - \Gamma_2 V_3)F_3(E)|^2 \rho_2(E) \\[1em] &\qquad\qquad + |\Gamma_3 + (\Gamma_2 V_{23} - \Gamma_3 V_2)F_2(E)|^2 \rho_3(E)\}, \qquad (6.1) \end{aligned}$$

128

where $E = ck - ck'$ is the energy transfer.

I discuss the irregularity of the Raman spectrum at E_c taking the example at $E_c = 3\Delta_0$. The transition probability, to which the Raman intensity is proportional, contains a term directly proportional to $\rho_3(E)$ which has a term $\sqrt{E - 3\Delta_0}$ above $3\Delta_0$. Moreover, $F_3(E)$, and consequently $D(E)$, also has the irregular term. By writing the definition (5.11) as

$$F_3(E) = \frac{1}{3!(2\pi)^6} \int \frac{1}{E - \omega_3(\vec{q},\vec{q}') + i\varepsilon} \, d\vec{q}\,d\vec{q}' \qquad (6.2)$$

and applying a formula $1/(x + i\varepsilon) = P(1/x) - i\pi\delta(x)$, it is found that $\text{Im}\{F_3(E)\}$ is just $-\pi\rho_3(E)$, which has $\sqrt{E - 3\Delta_0}$ above $3\Delta_0$. Because $F_3(E)$ is analytic in the upper half of the complex E-plane, $\text{Re}\{F_3(E)\}$ necessarily has $\sqrt{3\Delta_0 - E}$ below $3\Delta_0$. Therefore,

$$F_3(E) = F_3(3\Delta_0) + a_{3\Delta_0}\{\sqrt{3\Delta_0 - E}\,\theta(3\Delta_0 - E) - i\sqrt{E - 3\Delta_0}\,\theta(E - 3\Delta_0)\}$$

$$+ O(E - 3\Delta_0). \qquad (6.3)$$

$F_3(E)$ behaves similarly at the other $E_c = 2\Delta_0 + \Delta_1$, $\Delta_0 + 2\Delta_1$, $3\Delta_1$. By inserting (6.3) into (5.13) and (6.1), simple algebra shows that the Raman spectrum $I(E)$ behaves near each E_c as

$$I(E) = I(E_c) + A_c\sqrt{E_c - E}\,\theta(E_c - E) + B_c\sqrt{E - E_c}\,\theta(E - E_c) + O(E - E_c). \qquad (6.4)$$

The constants, A_c and B_c, depend on the value of Γ and V.

In this way I reached the conclusion with HIRASHIMA [2] that the Raman spectrum should have a cusp or inflection with infinite derivative just at E_c. Figure 6 shows the four possible types of the spectrum according to the sign of A_c and B_c.

The behavior of $F_2(E)$ near $2\Delta_0$ and $2\Delta_1$ can be deduced from the same argument. For example, I have near $2\Delta_0$,

$$F_2(E) = - a_{2\Delta_0}\left\{\frac{\theta(2\Delta_0 - E)}{\sqrt{2\Delta_0 - E}} + i\,\frac{\theta(E - 2\Delta_0)}{\sqrt{E - 2\Delta_0}}\right\} + C_{2\Delta_0} + O(E - 2\Delta_0)^{1/2}. \qquad (6.5)$$

Simple algebra shows that the expression (6.4) is also valid at $E_c = 2\Delta_0$ and $2\Delta_1$. The resonance energy E_r is determined by $\text{Re}\{D(E_r)\} = 0$. These points were already discussed in the references [3,4] without three-excitation corrections.

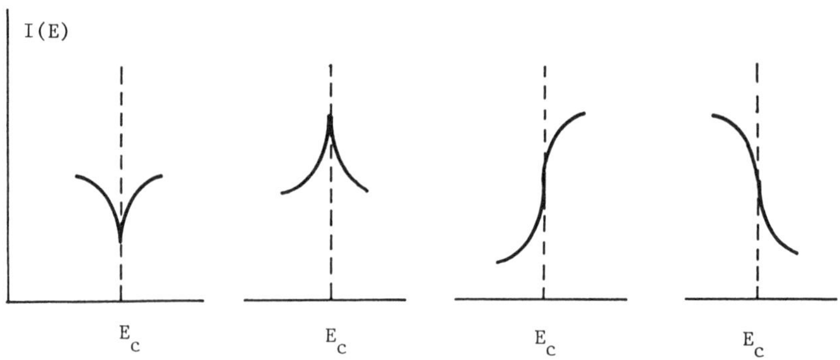

Fig.6 Four types of I(E) at $E_c = 3\Delta_0$, $2\Delta_0 + \Delta_1$, $\Delta_0 + 2\Delta_1$, $3\Delta_1$; $2\Delta_0$, $2\Delta_1$

Finally, I discuss a question about the uniqueness of Γ and V. I introduced in Sect.2 a set of Bose operators $b_{\vec{p}}$, $b_{\vec{p}}^{\dagger}$ so that $b_{\vec{p}}^{\dagger}|0>$ is the exact one-excitation state of momentum $\vec{p}$. However, this condition alone cannot fix the set uniquely. There are a variety of unitary operators U in the space of state vectors such that the ground state and the one-excitation states are left unchanged, $U|0> = |0>$, $Ub_{\vec{p}}^{\dagger}|0> = b_{\vec{p}}^{\dagger}|0>$, but $U|\alpha> \neq |\alpha>$. A new set of Bose operators $b'_{\vec{p}} = Ub_{\vec{p}}U^{-1}$, $b'^{\dagger}_{\vec{p}} = Ub_{\vec{p}}^{\dagger}U^{-1}$ gives Γ' different from Γ; $<0|b'_{-\vec{q}}b'_{\vec{q}}\Theta|0> \neq <0|b_{-\vec{q}}b_{\vec{q}}\Theta|0>$, etc. Also, the new interactions V' are different from V. For a further restriction, the asymptotic condition [15] will be suitable for defining the set. If any ambiguity still remains even under the asymptotic restriction, it does not affect the matrix M, combination of Γ and V, and the conclusion reached above is certainly valid as long as there exists at least one representation in which Γ and V are smooth.

The theory developed here is almost of kinematical nature. It is directly based on such fundamental laws as statistics, causality, time reversal, conservation of energy, momentum, angular momentum, parity and on certain reasonable assumptions. Therefore, the conclusion must be reliable.

References

1. K. Ohbayashi and M. Udagawa, Phys.Rev. B31 (1985) 1324.

2. D. S. Hirashima and F. Iwamoto, Prog.Theor.Phys. 75 (1986) 744.

3. F. Iwamoto, Prog.Theor.Phys. 44 (1970) 1135.

4. J. Ruvalds and A. Zawadowski, Phys.Rev.Lett. 25 (1970) 333.
    A. Zawadowski, J. Ruvalds and J. Solana, Phys.Rev. A5 (1972) 399.

5. L. D. Landau, J.Phys.USSR $\underline{5}$ (1941) 71; $\underline{11}$ (1947) 91.

6. E. P. Wigner, Phys.Rev. $\underline{73}$ (1948) 1002.

7. L. P. Pitaevskii, Sov.Phys.JETP $\underline{9}$ (1959) 830.

8. J. W. Halley, Phys.Rev. $\underline{181}$ (1969) 338.

9. T. J. Greytak and J. Yan, Phys.Rev.Lett. $\underline{22}$ (1969) 987.

10. A. L. Fetter,Phys.Rev. $\underline{140}$ (1965) A1921.

11. C. A. Murray, R. L. Woerner and T. J. Greytak, J. of Phys. $\underline{C8}$ (1975) L90.

12. R. L. Woerner and T. J. Greytak, J.Low Temp.Phys. $\underline{13}$ (1973) 149.

13. L. P. Pitaevskii and I. A. Fomin, Sov.Phys.JETP $\underline{38}$ (1974) 1257.

14. K. M. Watson, Phys.Rev. $\underline{88}$ (1952) 1163.

15. H. Lehmann, K. Symanzik and W. Zimmermann, Nuovo Cimento $\underline{1}$ (1955) 205.

Reflections on Excitations in Superfluid Helium

J. Ruvalds

Physics Department, University of Virginia, Charlottesville, VA 22901, USA

A brief review of the excitations in liquid He[4] is presented, with an emphasis on the interactions which were found to give two-roton resonances. The implications of this discovery were particularly relevant to the analysis of the Raman spectrum and neutron scattering experimental data. Later studies also demonstrated the inadequacy of the Born Approximation as a result of the resonant structure, and hence extensive theoretical progress was achieved by various groups in understanding the transport properties associated with phonons and rotons. Attempts to derive a truly microscopic description of the roton coupling encountered difficulties which were similar to those afflicting the meson scattering problem in nuclear physics. Nevertheless, semi-phenomelogical inputs for the structure factor led to successful descriptions of the lowest order interactions, although the details of the resonances remain beyond the reach of current theory. Recent developments in the study of roton interactions as a function of pressure suggest directions for future studies.

I. Phonons, Rotons, and Bound States

The elementary excitation spectrum of liquid helium has provided a number of theoretical challenges in the last few decades. At the inception by Landau[1] in 1947, it was recognized that thermodynamic properties such as the specific heat and viscosity could be explained by contributions from ordinary acoustic phonons at low momentum and rotons corresponding to an energy gap Δ_0 at intermediate momentum. The roton dispersion is commonly approximated by the expansion

$$E(k) = \Delta_0 + \frac{(k-k_0)^2}{2\mu_0} \quad , \tag{1}$$

in the vicinity of the roton minimum where $k \simeq k_0$, and a parabolic dispersion with effective mass μ_0 describes the thermodynamic data quite well at low temperatures. At higher energies, the excitation spectrum becomes far more complex and challenging from the theoretical point of view. The overall spectrum is shown in Fig. 1.

Microscopic theories of a dilute Bose gas yield the continuous phonon-roton spectrum as demonstrated originally by Bogoliubov,[2] providing that there is a substantial Bose condensate, and the interatomic potential in the liquid is of a certain shape. A classic alternate derivation of the roton region was found by Feynman and Cohen, who relied on the liquid structure factor in lieu of the unknown interatomic potential.[3] Their results also yield a continuous phonon-roton curve shown by the dashed curve in Fig. 1, but their estimate of the roton energy is somewhat higher than the experimental value.

Springer Series in Solid-State Sciences, Vol. 79 **Elementary Excitations in Quantum Fluids**
Editors: K. Ohbayashi · M. Watabe

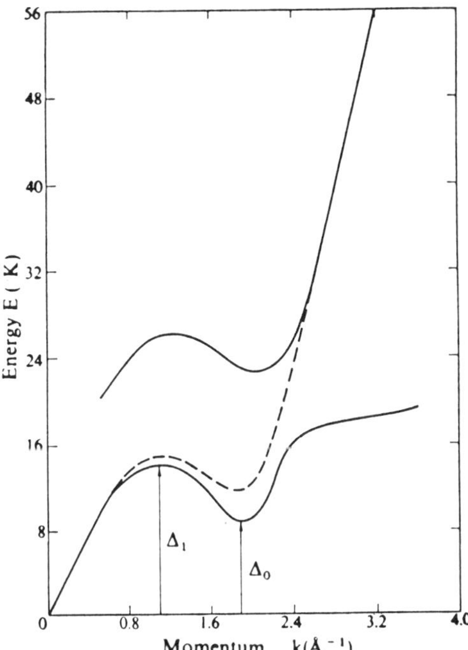

Figure 1. The excitation spectrum of superfluid helium deduced by Landau and derived by Feynman and Cohen is the continuous dashed curve. Neutron scattering data yields the two branch spectrum shown by the solid curve.

A direct measurement of the roton region by inelastic neutron scattering[4] revealed the complex structure in the excitations shown by the solid curve in Fig. 1, and attributed the higher energy branch near $E \simeq 2\Delta_0$ to "multiphonon" processes by virtue of the very broad structure in this region.[4]

Interest in the interactions among rotons in particular evolved rapidly in 1970 following the suggestions that a pair of rotons may form one or more bound states in various angular momentum channels. The initial motivation for this idea came from the Raman spectrum of liquid He[4] obtained by Greytak et al.[5] They found an asymmetric peak near the two-roton energy $2\Delta_0$ as expected, since the momentum of the light is so small that two rotons with total momentum $\underset{\sim}{K} = \underset{\sim}{k}_1 + \underset{\sim}{k}_2 \simeq 0$ are created in the

Raman process. Surprisingly, however, the Raman data failed to show a peak corresponding to the two-"maxon" region $E \simeq 2\Delta_1$, even though these "maxon" excitations yield a singular density of states in league with the rotons by virtue of a vanishing group velocity near the roton minimum or, alternately, near the "maxon" region shown by $k = k_1$ in Fig. 1. This puzzling feature stimulated our interest at essentially the same time that very similar conclusions were evolving in Japan by different techniques. Our approach[6,7] emphasized the singularity in the two-roton density of states which is of the form

$$\rho_2^0 (K=0,\omega) \simeq \left[\frac{k_0}{2\pi}\right]^2 \left[\frac{\mu_0}{\omega-2\Delta_0}\right]^{1/2} + \left[\frac{k_1}{2\pi}\right]^2 \left[\frac{\mu_1}{2\Delta_1-\omega}\right]^{1/2} \quad , \tag{2}$$

which illustrates the symmetry of roton and maxon contributions. Thus, we deduced that an attractive interaction between these excitations would yield a bound state of two rotons for arbitrarily weak coupling because even a minimal shift of the two-roton branch to lower energies yields a pole in the scattering amplitude by virtue of the singular density of states near $2\Delta_0$. By contrast, an attraction between maxons would shift their singular peak to lower energies within the continuum of two-maxon states, thus removing the singular character of the density of states.[6,7] Before demonstrating this phenomena more precisely, we present the alternate physical arguments advanced by Iwamoto,[8] which lead to similar conclusions. He noted that a roton excitation with energy near the minimum Δ_0 has zero group velocity despite having a finite momentum k_0, and therefore the Born Approximation for the scattering breaks down and resonances are likely to be formed in various angular momentum channels. The view of a roton as a cloud with a helium particle at its core and a backflow of neighboring atoms helps to visualize this phenomena. Thus, by solving the multiple scattering problem for two rotons using a separable potential, Iwamoto independently found the criterion for creating bound states of two excitations.

The theoretical two-roton spectrum which results from these calculations [6,7] is shown in Fig. 2.

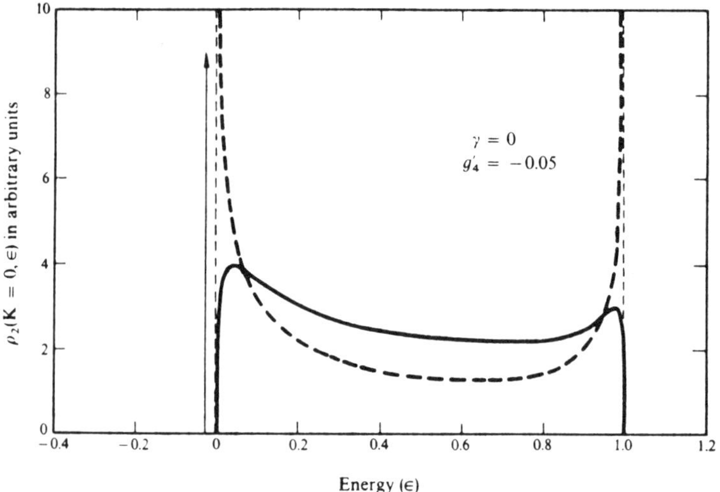

Energy (ϵ)

Fig. 2. Calculated density of states for two rotons with zero total momentum, plotted as a function of dimensionless energy $\epsilon = (\omega - 2\Delta_0)/2D$. Dashed curves show the spectrum without interactions, whereas the bound state formation is indicated by the arrow and the renormalized spectrum follows the solid curve for an attractive roton-roton coupling $g'_4 = -0.05$.

As seen clearly in Fig. 2, the influence of a weak interaction between excitations g'_4 (defined in Ref. 7) is to remove both singularities in the spectrum at $2\Delta_0$ and $2\Delta_1$, and also to produce a bound state of two rotons at $2\Delta_0 - F_B$ for an <u>attractive</u> coupling. Of course, there are many qualifications to keep in mind regarding the details of the observed Raman spectrum. First of all, the Raman polarization experiments indicate that the prominent peak near $2\Delta_0$ is of d-like angular momentum, and then the behavior of other angular momentum channels should be studied independently. Regarding the maxon branch, it would be somewhat fortuitous if they were coupled by the same interaction which influences the rotons, so that the absence of a two-maxon peak in the Raman spectrum suggests only a maxon-maxon attraction in the same $\ell = 2$ channel. Finally, the detailed line shape of the spectrum is valid only near the threshold $2\Delta_0$ by virtue of the effective mass approximation given in Eq.1 and used in the calculations. At energies far above $2\Delta_0$ (or far removed from $2\Delta_1$) higher order corrections to the dispersion must be included to achieve a quantitative analysis.

After considerable controversy, amid thorough studies in various laboratories, the highly refined Raman experiments were able to prove that a bound state in fact exists with a binding energy $E_B \simeq 0.4$ K in the d-wave channel.[9]

Two rotons with finite total momentum K may also be bound by an attractive interaction but in this case the prospects are dimmed by several factors. A principal feature of the finite momentum is a modification of the phase space available to a pair of scattering rotons, which yields a threshold in their density of states at $\omega = 2\Delta_0$ and also at $2\Delta_1$ for the maxon pair.[7] Nevertheless an attractive coupling will yield a bound state of two rotons at finite momentum, but the binding becomes progressively weaker as K increases. Hence the additional constraints of collision broadening and hybridization of these two-roton states with the single particle continuum compounds the experimental difficulties in observing these modes.

Hybridization of the single excitation branch with the two-roton continuum accounts for the extended plateau in the spectrum seen in Fig. 1 near $\omega \simeq 2\Delta_0$, and this process was originally proposed by Pitaevskii.[10] Our efforts extended this concept to include final state interactions in the scattering of two rotons, by analogy to the Fermi resonance ideas in atomic physics which allow the observation of an overtone of an atomic level in a first order process by virtue of the level repulsion between the overtone and a single level. Although the final state interactions among the rotons appeared to improve the agreement with neutron scattering data in the multiphonon regime, i.e. $\omega > 2\Delta_0$, the substantial broadening in these experiments prevented a clear-cut identification of a narrow two-roton resonance at finite momentum.[7]

On the other hand, estimates of the roton coupling parameters were more than an order of magnitude lower than the values previously used in the context of the Born approximation to explain a variety of transport data. This discrepancy stimulated intense activity in the multiple scattering theory of the helium excitations.

II. Energy and Lifetime Calculations

In view of the resonant scattering features for two rotons, the validity of the Born approximation in treatments[11] of transport properties was reexamined by several groups.[12−15] Remarkably they all reached similar but unexpected conclusions. If the roton-roton coupling is approximated by a simple contact interaction (equivalent to a δ-function potential in real space), the multiple scattering terms in the scattering amplitude make it impossible to reconcile the value of the roton binding energy with the roton lifetime data. Although this dilemma is resolved by the observation that the Raman active state is of d-wave symmetry, and a contact potential represents s-wave scattering, the more fundamental difficulty of achieving a good description of the observed lifetime requires scattering in several angular momentum channels.

Physically, the reason for the above complexities can be seen from the expression for the roton lifetime [14] τ,

$$\frac{1}{\tau} = \frac{g_4^2}{2\pi^3} \int d^3K' \, \rho_2 \, (\vec{K}+\vec{K'}, \, \omega+E_{K'}) \, n_B(E_K') \quad , \tag{3}$$

where n_B is the Bose distribution function and E_K is the roton energy. Starting with the Born approximation and an energy dependent density of final scattering states ρ_2^0 $(\vec{K} \neq 0)$, it is clear that a value of the coupling g_4 can be chosen to achieve any value of τ. However, the final state interactions yield a dramatic change in the density of states, which becomes

$$\rho_2 (K,\omega) = - \frac{1}{2\pi} \, \text{Im} \, \frac{F(K,\omega)}{1-g_4 F(K,\omega)} \quad , \tag{4}$$

where

$$F(K,\omega) = 2 \int_0^{2D} \frac{\rho_2^{(0)}(K,\omega')d\omega'}{\omega-\omega'+i\Gamma} \quad . \tag{5}$$

The fact that ρ_2 in Eq. 4 becomes dependent on the coupling g_4 imposes a unitarity limit on the scattering cross section (or τ^{-1}) and it turns out that this maximum value in a given angular momentum channel is insufficient to reproduce the experimental value of τ. Naturally, the combined contributions of several angular momentum channels will suffice, but then the theory must rely on several coupling parameters.

Softening of the roton energy with temperature provides an alternate challenge for the theory, especially since this phenomena is closely connected to the depletion of the superfluid density and is of particular interest near the λ phase transition.

Again the multiple scattering corrections associated with the two-roton resonance become important, and the calculations extending beyond the Hartree-Fock approximation reveal significant corrections. The temperature variation of the roton minimum is shown in Fig. 3.

More refined calculations of the roton energy [17-19] yield another constraint on the roton-roton coupling, and they suggest that seven or more angular momentum channels should exhibit a strong attractive component in order to achieve agreement with the observed Δ (T) variation.

In view of the uncertainties in the evidently strong momentum variation of the roton-roton coupling, efforts evolved to obtain an empirical pseudopotential and also to develop a microscopic theory of these interactions.

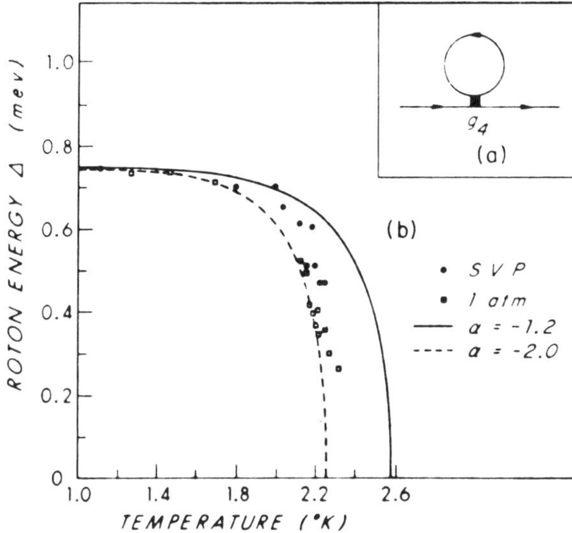

Figure. 3   The calculated roton energy from Ref. 16 is compared to experimental data at two pressures over a wide range of temperatures.

III. Microscopic Theories of Roton Coupling

A first principles description of the Bose gas has been the object of numerous studies over the last thirty years. Ultimately the goal of achieving a reasonable description of the phonon-roton excitations has been approached with good accuracy, but the residual interactions between quasiparticles remain unresolved. [20]

Here we shall not go into the details of the difficulties but rather point out the fundamental obstacles which the quantum liquids have in common with other fields of physics, such as the scattering of mesons in nuclei.

As a first approximation, the force between two rotons can be estimated by the familiar process of phonon exchange, and the magnitude of this term turns out to be reasonable. However, the exchange of a roton yields a coupling of the same order and, more crucially, the exchange of two rotons gives a scattering amplitude comparable in value.[20] Thus, no matter how sophisticated is the initial formulation of the microscopic Hamiltonian, the hierarchy of scattering terms extends to infinite order, with little hope of a convergent tractable solution.

In some ways the roton resonance difficulties are reminiscent of the meson exchange originally postulated by Yukawa. The ρ meson can be considered as a resonance in the scattering of two π mesons, and the exchange of a single π meson gives a reasonable force. However the exchange of a ρ meson in the $\pi - \pi$ scattering is also significant and thus a straightforward derivation of the ρ meson energy and lifetime is difficult to achieve using only the single π-meson spectrum as input.

Despite the above difficulties, phenomenological descriptions of the roton interactions have achieved considerable progress. One promising approach is an expansion of the Hamiltonian in terms of density variables, and, at some late stage in the analysis, the introduction of the experimental structure factor as input for the calculation of interaction vertices.[21,22] In some sense this corresponds to the introduction of a pseudopotential to represent the interatomic forces in the liquid state.

Other approaches have also been suggested to represent the helium pseudopotential in terms of adjustable parameters and then to deduce the true nature of the interactions from comparisons to experiments as a function of pressure as well as temperature.[23]

IV. Prospects for Future Studies

In the continuing quest to understand the excitations in superfluid helium, more refined neutron and light scattering experiments are perhaps the most promising. The recent polarization studies of the Raman spectrum [24] confirm the fundamental issues such as the existence of a two-roton bound state in the d-wave angular momentum channel and argue against a significant peak in the s-wave channel. By applying pressure, these studies demostrate a shift in the two-roton peak which can be correlated with the single roton energy as a function of pressure obtained from neutron data. These experiments present some interesting theoretical questions regarding the sign and magnitude of the roton interaction under pressure.

Furthermore, further studies at even higher resolution may reveal additional structure in the multi-excitation regime.

This work has been supported by the U.S. Department of Energy, Grant Number: DEFGO584ER45113

REFERENCES

1. L. D. Landau, J. Phys. USSR $\underline{11}$, 91 (1947).
2. N.N. Bogoliubov, J. Phys. USSR $\underline{11}$, 23 (1947).
3. R. P. Feynman and M. Cohen, Phys. Rev. $\underline{102}$, 1189 (1956).
4. R.A. Cowley and A.D.B. Woods, Can. J. Phys. $\underline{49}$, 177 (1970).
5. T.J. Greytak, R. Woerner, J. Yan and R. Benjamin, Phys. Rev. Lett. $\underline{25}$, 1547 (1970).
6. J. Ruvalds and A. Zawadowski, Phys. Rev. Lett. $\underline{25}$, 333 (1970).
7. A. Zawadowski, J. Ruvalds and J. Solana, Phys. Rev. $\underline{A5}$, 399 (1972).
8. F. Iwamoto, Prog. Theor. Phys. (Kyoto) $\underline{44}$, 1135 (1970).
9. C.A. Murray, R. L. Woerner and T.J. Greytak, J. Phys. $\underline{C8}$, L90 (1975).
10. L.P. Pitaevskii, Zh. Eksp. Teor. Fiz. $\underline{36}$, 1168 (1959).
11. L.D. Landau and I.M. Khalatnikov, in Collected Papers of L.D. Landau (ed. D. Ter Haar, Gordon & Breach, NY, 1967), pp. 494,511.
12. J. Yau and M. Stephen, Phys. Rev. Lett. $\underline{27}$, 482 (1971).
13. I.A. Fomin, Sov. Phys. JETP $\underline{33}$, 637 (1971).
14. J. Solana, V. Celli, J. Ruvalds, I. Tutto and A. Zawadowski, Phys. Rev. $\underline{A6}$, 1665 (1972).
15. K. Nagai, K. Nojima and A. Hatano, Prog. Theo. Phys. (Kyoto) $\underline{47}$, 355 (1972).
16. J. Ruvalds, Phys. Rev. Lett. $\underline{27}$, 1769 (1971).
17. K. Nagai, Prog. Theor. Phys. $\underline{49}$, 46 (1973).

18. I. Tüttö, J. Low Temp. Phys. $\underline{11}$, 77 (1973).
19. T. Kebukawa, Prog. Theor. Phys. $\underline{49}$, 388 (1973).
20. A.K. Rajagopal, A. Bagchi and J. Ruvalds, Phys. Rev. $\underline{A9}$, 2707 (1974); this article contains an extensive list of references to previous work.
21. A.K. Rajagopal and G.S. Grest, Phys. Rev. $\underline{A10}$, 1837 (1974); ibid, 1395 (1974).
22. J.A. Carballo and J. Ruvalds, Phys. Rev. $\underline{B11}$, 4278 (1975).
23. K.S. Bedell, D. Pines and A. Zawadowski, Phys. Rev. $\underline{B29}$, 102 (1984).
24. M. Udagawa, H. Nakamura, M. Murakami and K. Ohbayashi, Phys. Rev. $\underline{B34}$, 1563 (1986).

On Pitaevskii's Branch near the Two-Roton Threshold at Finite Temperatures

Katsuhiko Nagai

Department of Physics, Yamaguchi University, Yamaguchi 753, Japan

PITAEVSKII[1] predicted in 1959 that the phonon-roton curve will terminate tangentially to the edge of the two-roton continuum at certain momentum Q_c. In the neutron experiments, however, this asymptotic branch by PITAEVSKII has not been observed up to the end point Q_c. The peak position is found to be at rather higher energy than twice the roton energy at large momentum transfer. This report shows that, at finite temperatures, due to the finite linewidth of rotons PITAEVSKII's branch ceases to exist at a temperature dependent momentum $Q_c(T)$ which is smaller than PITAEVSKII's Q_c, thereby explaining the above discrepancy. A brief report has been given by NAGAI[2]. ZAWADOWSKI et al.[3] have also shown numerically that the finite linewidth of rotons shifts PITAEVSKII's branch to higher energy.

Let us first briefly review PITAEVSKII's theory and extend it to finite temperatures. The Green function G of the elementary excitation obeys the Dyson equation

$$G^{-1}(p) = G_0^{-1}(p) - \Sigma(p),\tag{1}$$

where $p=(Q,\omega)$ is the 4-dimensional energy momentum, G_0 is the unperturbed Green function and Σ the self energy. In the present problem, the three particle process is important; therefore the self energy may be written as

$$\Sigma(p) = \int g_3(p;q,p-q)G(q)G(p-q)\Gamma_3(p;q,p-q)\ d^4q/(2\pi)^4,\tag{2}$$

where g_3 is the unperturbed three-particle vertex and Γ_3 is the complete vertex. Near the 2Δ threshold, Γ_3 has a singularity which comes from the 2-roton propagator. So we rewrite Γ_3 by explicitly extracting the 2-roton propagator and by putting the other nonsingular parts into the irreducible vertex g_4:

$$\Gamma_3(p;q,p-q) = g_3(p;q,p-q)$$
$$+ \int g_4(q,p-q;r,p-r)G(r)G(p-r)\Gamma_3(p;r,p-r)\ d^4r/(2\pi)^4.\tag{3}$$

Near the two-roton threshold , g_3 and g_4 may be taken as constant; g_4 can be interpreted as the effective roton-roton coupling. Since the dominant contribution in the integrals of eqs. (2) and (3) comes from the two-roton region, the Green function of the internal lines can be put equal to the observed roton propagator

$$G^{-1}(p) = \omega - E(Q) + i\Gamma/2,\tag{4}$$

where $E(Q)$ is the observed roton energy

$$E(Q) = \Delta + (Q-Q_0)^2/2\mu\tag{5}$$

and Γ is the observed roton full linewidth which can be well described by Landau-Khalatnikov theory as

$$\Gamma = A \exp[-\Delta/T]. \tag{6}$$

Now one can solve the Dyson equation to have

$$\Sigma(Q,\omega) = (2g_3^2/g_4) \, g_4\rho F/(1-g_4\rho F), \tag{7}$$

where $\rho=Q_0^2\mu/2\pi Q$ the two roton density of states and F is

$$F(Q,\omega) = \ln\left|\sqrt{\tilde{\omega}^2+\Gamma^2}\,/2D\right| - i(\pi/2 + \tan^{-1}(\tilde{\omega}/\Gamma)\,), \tag{8}$$

where $\tilde{\omega}=\omega-2\Delta$ and D is the cut off energy of interacting rotons. It is a crucial point of the present theory that the roton linewidth is included in F, since the logarithmic singularity which appears at T=0K is easily smeared by tho finite linewidth even if Γ is so small as 10^{-2}K.

In what follows, we assume that the effective roton-roton coupling g_4 is repulsive at large total pair momentum Q, because it is compatible with the experiments as can be seen below. The pole of the Green function is given by

$$\omega - E_0(Q) - \Sigma(Q,\omega) = 0 \tag{9}$$

with $E_0(Q)$ the unperturbed excitation energy.

Let us first consider the case at T=0, therefore $\Gamma=0$. Near the threshold, the real part of Σ has a cusp minimum as

$$\Sigma(Q,\omega) = -(2g_3^2/g_4)\{\, 1 + (g_4\rho\ln|\tilde{\omega}/2D|)^{-1} + \cdots\, \}\,. \tag{10}$$

Since $E_0(Q)$ is expected to be an increasing function at large Q, one can find a critical momentum $Q_c(0)$ given by

$$2\Delta - E_0(Q_c(0)) = -2g_3^2/g_4\,. \tag{11}$$

For $Q < Q_c(0)$, eq.(9) has a solution below the 2-roton continuum, i.e.,

$$\omega - 2\Delta = -\alpha \exp[-a/(Q_c(0)-Q)], \tag{12}$$

which is nothing but PITAEVSKII's branch. The coefficients are given by

$$\alpha = 2D \exp[1/g_4\rho], \qquad a = 2(g_3/g_4)^2/\{\rho(\partial E_0/\partial Q_c(0))\}\,.$$

At finite temperatures, the cusp minimum of Σ is smeared only within a narrow range of $|\tilde{\omega}|<\Gamma$ ($\Gamma\sim10^{-2}$K at T=1.1K). One can find a critical momentum $Q_c(T)$ determined by

$$2\Delta - E_0(Q_c(T)) = (2g_3^2/g_4) \, g_4\rho\ln(\Gamma/2D)/(1-g_4\rho\ln(\Gamma/2D)). \tag{13}$$

Subtracting eq.(13) from eq.(11) and using the expression (6), one finds that

$$Q_c(T)-Q_c(0) = - a \ (T/\Delta) \ / \ \{1 + (T/\Delta)\ln(\alpha/A)\} \ . \tag{14}$$

For $Q<Q_c(T)$ and $\omega<2\Delta-\Gamma$, eq.(9) has a solution essentially the same as eq.(12) for T=0K, because the smearing of Σ occurs only within the narrow range of $|\tilde{\omega}|<\Gamma$. For $Q>Q_c(T)$, PITAEVSKII's branch entirely disappears and there remains only the 2-roton continuum. This agrees very well with the neutron scattering by GRAF et al.[4]. In fact, they found that the phonon-roton curve in the range $2.3\text{Å}^{-1}<Q<2.8\text{Å}^{-1}$ at SVP can be well fitted by eq.(12) and obtained $\alpha=1.2\times10^5$K, a=7.2 Å^{-1} and $Q_c(0)=3.55\text{Å}^{-1}$. Putting these parameters as well as $A\simeq100$K from the linewidth data(DIETRICH et al.[5]) into eq.(14), one can estimate $Q_c(T)$ at T=1.1K to be 2.85Å^{-1}.

For more quantitative discussion, it is convenient to compute the spectral weight ImG. For that purpose, we need information on the unperturbed excitation energy $E_0(Q)$. At such large momentum as $Q_c(0)$, however, it is plausible that $E_0(Q)$ will coincide with the energy of the ^{4}He atom free particle motion. Thus we may put

$$E_0(Q) = Q^2/2m_{^4He} \simeq 6.0\times Q^2 \ [\text{KÅ}^{-2}] \ . \tag{15}$$

Then, from eq.(11) together with the values of α and a, we find

$$g_3^2/g_4 \simeq 30[K], \quad 2D\simeq8[K], \quad g_4\rho\simeq0.2 \ ,$$

within the accuracy of single digit. In Fig.1, ImG at T=1.1K is plotted as a function of $\tilde{\omega}$ for Q=2.8, 3.0 and 3.2Å^{-1}. One can clearly see that PITAEVSKII's branch has disappeared at $Q=3.2\text{Å}^{-1}$. In Fig.2, ImG at T=0.9K is plotted. At this temperature, even at $Q=3.2\text{Å}^{-1}$ one can find a weak peak just at $\omega=2\Delta$.

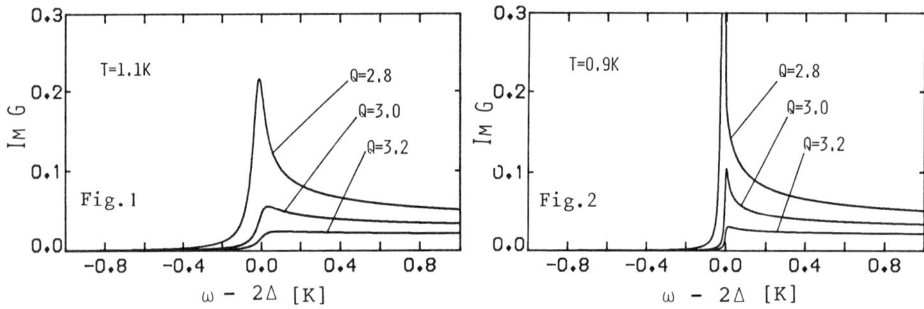

Fig.1　Im G is plotted versus energy at T=1.1K for various momenta.

Fig.2　Im G is plotted versus energy at T=0.9K for various momenta.

In a summary, we have shown that there is a temperature dependent $Q_c(T)$ at which PITAEVSKII's branch ceases to exist. The temperature dependence is very strong as inferred from eq.(14). This can be checked most easily by Raman scattering at the energy range of 2-plateau. Since the plateau (PITAEVSKII's branch) has a large density of states, the growth of the existence range in momentum space with decreasing temperature should be reflected in Raman scattering intensity.

Acknowledgement

The author expresses his thanks to Prof. K. Ohbayashi for stimulating
discussion on the elementary excitations in liquid He.

References

1. L.P. Pitaevskii: Sov. Phys. JETP $\underline{9}$, 830 (1959).
2. K. Nagai: J. Phys. C: Solid St. Phys. $\underline{11}$, L759 (1978).
3. A. Zawadowski, J. Ruvalds and J. Solana; Phys. Rev. $\underline{A5}$, 399 (1972).
4. E.H. Graf, V.J. Minkiewicz, H. Bjerrum Møller and L. Passell;
    Phys. Rev. $\underline{A10}$, 1748 (1974).
5. O.W. Dietrich, E.H. Graf, C.H. Huang and L. Passell:
    Phys. Rev. $\underline{A5}$, 1377 (1972).

Determination of the Roton-Maxon Interaction

Kenji Fukushima and Fumikazu Iseki

Department of Physics, Tokyo Metropolitan University, Tokyo 158, Japan

1. Introduction

It is remarkable that the peak of the upper branch of the excitation spectrum of liquid HeII discovered by Cowley and Woods /1/ lies between $2\Delta_0$ and $2\Delta_1$ for $0.5A^{-1} < |\vec{k}| < 2.0A^{-1}$; an energy region characteristic of two-quasiparticle excitations. Δ_0 and Δ_1 are, respectively, energies of roton and maxon. For $|\vec{k}| > 2.0A^{-1}$, as is well-known, it behaves like a free-particle excitation. On the other hand, the very broad upper branch has been looked upon as the contribution of multi-quasiparticle excitations.

Recently Manousakis and Pandharipande /2/ numerically calculated the two-quasiparticle excitation part, which will be written in this paper as $S_{II}(k)$, of the dynamical structure function, $S(k)$, by taking account of mixing of one- and two-quasiparticle excitations. They seem, however, not to succeed in reproducing the profile of neutron spectra near the peak of the upper branch for $0.5A^{-1} < |\vec{k}| < 2.0A^{-1}$. Here and hereafter k will stand for both wave vector $\vec{k}$ and frequency ω.

Fukushima, Koyama and Sugiyama /3/ once pointed out the possibility that the peak of the upper branch lying between $2\Delta_0$ and $2\Delta_1$ might be a resonance of roton-maxon pairs. By investigating in detail the $S_{II}(k)$, we in this paper first derive the condition for the existence of a resonance of roton-maxon pairs and then identify the above-mentioned peak of the upper branch as the roton-maxon resonance. With this identification, the sign and the strength of the roton-maxon interaction are determined, which turns out to be of the same order of magnitude as that of the effective roton-roton interaction. Consequently the roton-maxon interaction might play an important role in thermostatic and transport phenomena of liquid HeII.

2. Formulation

Let us confine ourselves to $S_{II}(k)$, assuming that three- and more-quasiparticle excitation parts of the $S(k)$ constitute a background against the remarkable structure around $\omega = \Delta_0 + \Delta_1$ in neutron spectra. Moreover, since relative values of $S_{II}(k)$ only are required for a comparison with neutron spectra, one may start /4/ from

144   Springer Series in Solid-State Sciences, Vol. 79   **Elementary Excitations in Quantum Fluids**
Editors: K. Ohbayashi · M. Watabe   © Springer-Verlag Berlin, Heidelberg 1989

$$S_{II}(k) \backsim -(1/\pi)\sum_{p,p'}\mathrm{Im}\, K(p,p';k), \tag{1}$$

where $K(p,p';k)$ is a two-particle propagator, p and p' being relative four-momenta. Formally $K(p,p';k)$ is defined as

$$K = 2\, K^0 / (1 - (1/2)\, g\, K^0), \tag{2}$$

where $K^0(p;k)$ is a free two-particle propagator, $g(p,p';k)$ being an effective two-quasiparticle interaction. Let us further restrict to such two-quasiparticle excitations as two-roton, roton-maxon and two-maxon, supposing that other pairs contribute a structureless background to the upper branch of the excitation spectrum. With this restriction and by only retaining the on-energy-shell component of the $g(p,p';k)$, may reduce (1) to

$$S_{II}(k) \backsim \left(\sum_i \rho_i(k)\right) / \left\{ \left(1 - \sum_i (g_i(\vec{k})/g_{ci}(\vec{k}))f_i(\omega)\right)^2 \right.$$
$$\left. + \left(\pi \sum_i g_i(\vec{k})\, \rho_i(k)\right)^2 \right\}, \tag{3}$$

where the subscript i refers to the kind of pairs. The $g_{rm}(\vec{k})$, for example, is the s-component /5/ of the effective roton-maxon interaction on the energy shell; $g_0(p_0\Delta_0, p_1\Delta_1; \vec{k}\Delta_0+\Delta_1)$. The densities of states of pairs are assumed to take the following forms:

$$\rho_{rr}(k) = \frac{\mu_0 p_0^2}{4\pi|\vec{k}|\hbar^2}\, \theta(\omega - 2\Delta_0)\, \theta(2\Delta_0 + \Delta - \omega),$$

$$\rho_{mm}(k) = \frac{\mu_1 p_1^2}{4\pi|\vec{k}|\hbar^2}\, \theta(2\Delta_1 - \omega)\, \theta(\omega - 2\Delta_1 + \Delta), \tag{4}$$

$$\rho_{rm}(k) = \frac{\sqrt{\mu_0\,\mu_1}\,p_0 p_1}{2\pi^2|\vec{k}|\hbar^2}\ln\frac{2\Delta}{|\omega - \Delta_0 - \Delta_1|},$$

where $\Delta = (\Delta_1 - \Delta_0)/2$. Suppose that $\rho_{rr}(k)$ and $\rho_{mm}(k)$ in (4) gradually decrease to zero toward $\Delta_0 + \Delta_1 + \Delta$ and toward $\Delta_0 + \Delta_1 - \Delta$, respectively. The free two-particle propagator integrated over p

$$\frac{1}{V}\sum_{\vec{p}}\int\frac{d\epsilon}{2\pi}(1/2)\, K^0(p;k) = \int d\epsilon'\, \frac{\rho_i(\vec{k}\,\epsilon')}{\epsilon - \epsilon' + i\delta}$$

may, by the use of (4), amount to

$$f_{rr}(\omega) = -\ln\frac{2\Delta}{|\omega - 2\Delta_0|},$$

$$f_{mm}(\omega) = \ln \frac{2\Delta}{|2\Delta_1 - \omega|} , \tag{5}$$

$$f_{rm}(\omega) = f((\omega - \Delta_0 - \Delta_1)/\Delta),$$

where $f(x) = (2/\pi^2) \int_0^{2/x} \frac{dt}{t} \ln \left| \frac{t+1}{t-1} \right|$, apart from such factors that $g_{crr}(\vec{k}) = 4\pi|\vec{k}|\hbar^2/\mu_0 p_0^2$, $g_{cmm}(\vec{k}) = 4\pi|\vec{k}|\hbar^2/\mu_1 p_1^2$ and $g_{crm}(\vec{k}) = 4|\vec{k}|\hbar^2/\sqrt{\mu_0\mu_1}\,p_0 p_1$. The $f(x)$ is an odd function of x, discontinuous at $x = 0$ with $f(+0) = -f(-0) = 1$ and monotonically decreasing to zero for $x > 0$.

3. Roton-Maxon Resonance

The condition for the occurrence of bound or resonant pairs is that the first term of the denominator on the r. h. s. of (3) should vanish. On the other hand, it has been shown in (4) and (5) that the singularities of the density of states of pairs, discontinuities at $\omega = 2\Delta_0$ and at $\omega = 2\Delta_1$, and logarithmic divergence at $\omega = \Delta_0 + \Delta_1$, result in a singular two-particle propagator, logarithmically divergent at $\omega = 2\Delta_0$ and at $\omega = 2\Delta_1$, and discontinuous at $\omega = \Delta_0 + \Delta_1$.

From the above-mentioned condition together with the singular functions (5), it immediately follows that any attractive $g_{rr}(\vec{k})$, however weak it is, leads to the formation of a bound roton-pair with an energy below $2\Delta_0$ and at the same time a resonant pair with an energy above $2\Delta_0$, and similarly any repulsive $g_{mm}(\vec{k})$ forms both a bound maxon-pair and a resonant one above and below $2\Delta_1$, respectively. Thus the singular density of states of pairs may be seen to play a role of a magnifying glass. The roton-maxon pair will be referred to later on.

Cowley and Woods /1/ showed profiles of neutron spectra for several values of $|\vec{k}|$, which seems to provide no simultaneous evidence of a sharp peak below $2\Delta_0$ and a broad one above $2\Delta_0$ and so on. These facts suggest a repulsive $g_{rr}(\vec{k})$ and an attractive $g_{mm}(\vec{k})$. Note that this conclusion is not decisive owing to the poor accuracy of the experiments.

In Fig.1, $\sum_i (g_i(\vec{k})/g_{ci}(\vec{k}))f_i(\omega)$ is depicted against ω for $|\vec{k}| = 1.0A^{-1}$, where parameters used are shown in the figure caption. The figure for negative $g_{rm}(\vec{k})$ may be immediately obtained from Fig.1 by reversing the direction of the abscissa and by exchanging parameters of roton and maxon. It may follow from Fig.1 and the figure for negative $g_{rm}(\vec{k})$ that if the absolute value of $g_{rm}(\vec{k})$ is greater than a critical strength $g_{crm}(\vec{k})$, there exists a resonance level of roton-maxon pairs, and that, depending on repulsion or attraction of

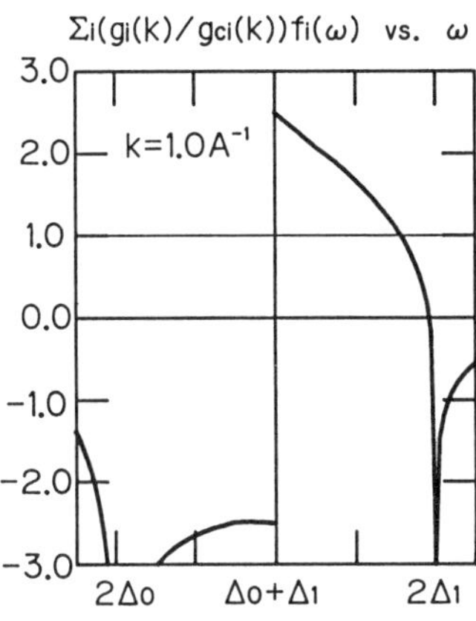

Fig.1. $\sum_i (g_i(\vec{k})/g_{ci}(\vec{k}))f_i(\omega)$ is depicted against ω for $|\vec{k}| = 1.0\text{Å}^{-1}$, where $g_{rr}(\vec{k})/g_{crr}(\vec{k}) = 1.0$, $g_{mm}(\vec{k})/g_{cmm}(\vec{k}) = -0.4$ and $g_{rm}(\vec{k})/g_{crm}(\vec{k}) = 2.5$ are used.

The intersection with a straight horizontal line passing through a point (0,1) defines a resonance energy.

$g_{rm}(\vec{k})$, it lies above or below $\Delta_0 + \Delta_1$. Here we want to note that owing to the strong ω-dependence of $\rho_{rm}(k)$ the resonance energy not always coincides with the solution of the afore-mentioned equation.

In Fig.2, we depicted $S_{II}(k)$ against ω for $|\vec{k}| = 1.0A^{-1}$ under the same values of parameters as those in Fig.1. At the same time the observed intensities are also depicted by suitably subtracting the background and by arbitrarily scaling the remainder. Cowley and Woods /1/ ascribed a single-peaked structure to the profiles of neutron spectra observed for several values of $|\vec{k}|$ and thereby it might be plausible to identify the peak as the resonance of roton-maxon pairs.

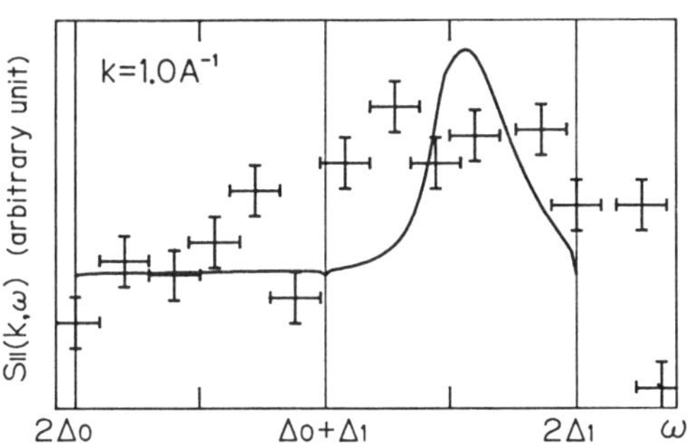

Fig.2. $S_{II}(k)$ is depicted against ω for $|\vec{k}| = 1.0\text{Å}^{-1}$, where the values of $g_i(\vec{k})/g_{ci}(\vec{k})$ are the same as those in Fig.1. The observed intensities /1/ are also depicted.

4. Roton-Maxon Interaction

A comparison with the observed position of peak of the upper branch
may lead to the following conclusion; $g_{rm}(\vec{k})$ is repulsive for $0.65A^{-1}$
$< |\vec{k}| < 2.0A^{-1}$, since the peak lies above $\Delta_0 + \Delta_1$, but it seems to
change sign for $|\vec{k}| < 0.65A^{-1}$. In particular, the strength of $g_{rm}(\vec{k})$
may be comparable with or at most a few times greater than $g_{crm}(\vec{k})$
around $|\vec{k}|$ $1.3A^{-1}$. Here the critical strength is

$$g_{crm}(\vec{k}) = |\vec{k}| \text{ (in } A^{-1}) \times 0.98 \times 10^{-38} \text{ erg cm}^3. \tag{6}$$

This value is almost of the same order of magnitude as that of the
effective roton-roton interaction /6/.

5. Discussion

So far we have ignored the finite width of roton and maxon, and the ω
-dependence and the off-energy-shell component of the effective two-
quasiparticle interaction. Moreover we have not estimated the
background. For a quantitative comparison and to draw a decisive
conclusion, it is required to refine the calculation by incorporating the above-mentioned points, as well as to   improve the
accuracy of experiments.

On the other hand, Svensson et al./7/ carried out detailed
measurements of the upper branch of neutron spectra and realized
that "the multiphonon component exhibits a complex shape with a
long high-frequency tail". With a new generation of high-resolution
neutron inelastic scattering instruments, Stirling/8/ has recently
reinvestigated the dispersion curve of elementary excitations and
the profile of the upper branch of neutron spectra. In particular,
the latter exhibited apparent multi-peaked structure within the
energy range between $2\Delta_0$ and $2\Delta_1$, which is very
sensitive to applied pressures and strongly depends on wave vector
transfers.

As Stirling has mentioned in his paper/8/, we too believe that
the observed structure and its dependence on pressure and wave
vector transfer should be explained by the excitation of the
quasiparticle-pair composed of interacting roton and maxon, by
taking account of the pressure-dependence of Δ_0 and Δ_1, and the
wavevector- and pressure-dependence of the effective two-quasiparticle
interaction. The mixing of the one- and the two-quasiparticle
excitations, which has already been taken into consideration by
Manousakis and Pandaripande/2/ should be also incorporated for such
high applied pressures that $2\Delta_0$ approaches Δ_1.

Finally we want to stress that the roton-maxon interaction should
play a significant role in understanding thermostatic and transport
phenomena of liquid HeII.

References

1. R.A. Cowley and A.D.B. Woods, Can. J. Phys. <u>49</u>, 177 (1971).
2. E. Manousakis and V.R. Pandharipande, Phys. Rev. B<u>33</u>, 150 (1986).
3. K. Fukushima, N. Koyama and T. Sugiyama, Prog. Theor. Phys. <u>61</u>, 367 (1979).
4. A. Zawadowski, J. Ruvalds and J. Solana, Phys. Rev. A<u>5</u>, 399 (1972).
5. I.A. Fomin, Soviet Physics JETP <u>33</u>, 637 (1971).
6. K. Bedell, D. Pines and A. Zawadowski, Phys. Rev. B<u>29</u>, 102 (1984) and references cited therein.
7. E.C. Svensson, P. Martel, V.F. Sears and A.D.B. Woods, Can. J. Phys. <u>54</u>, 2178 (1976).
8. W.G. Stirling, Invited paper, International Conference on Phonon Physics, Budapest, August 1985.

Two-Particle Excitations and Pair Correlations in He-II

R. Sridhar

Institute of Mathematical Sciences, T.T.T.I-Taramani, Madras 600113, India

1. Introduction

In this contribution we are concerned with the role of two-particle excitations in liquid He-II. This has become an important aspect to be understood by microscopic theories of the superfluid since Cowley and Woods /1/ reported in their neutron scattering intensities the observation of a broad distribution whose energy width and intensities are strong functions of the momentum transfer k. This peak was seen in addition to the rather sharp peak usually associated with one phonon scattering. The intensity of the broad maximum rises to about 25K and extends to relatively large(about 80K) energy transfers. Such a broad peak was not altogether an unexpected one. Phenomenological arguments of Miller, Pines and Nozieres /2/ indicate the existence of a broad background.

However, an explicit observation of this background led several theoreticians to postulate that this may indicate the existence of a rather well defined set of excitations in the liquid hitherto unobserved. These ideas gained a further fillip by the observation by Greytak and Yan /3/ of a sharp peak (in laser Raman scattering from liquid Helium) near twice the roton gap. It was reported that the peak occurred slightly below 2Δ which led to the conjecture of a two-roton bound state. This interpretation prompted a few objections. But, later repetitions of the experiment seem to have settled the issue in favour of two-roton bound state. At a time when the idea was not very actively pursued came the experimental report of Blagoveshchenskii et al /4/ regarding the observation, through neutron scattering, of two-roton bound states with non-zero total momentum. Also the recent Raman scattering results due to Ohbayashi and Udagawa /5/ have added several details to be explained.

2. Motivation

We mention a slightly apparently different problem-the rather surprising behaviour of the pair correlation function of He-II, observed both in neutron scattering and X-ray scattering /6,7/ studies. The principal peak of the liquid structure factor S(k) - which is, in a sense, the Fourier transform of the pair correlation function - become narrower and started increasing in height when the liquid was cooled from well above the lambda temperature T_λ. This clearly indicated reduction in the thermal motions resulting in greater local order. However, when the temperature is lowered still below T_λ, the principal peak started decreasing in height with an increase in its width as well as signalling a loss in spatial order.

It was earlier predicted by Hylands et al /8/ that Bose-Einstein condensation of He atoms into the zero momentum state is responsible for this loss in spatial order. But this idea was criticised on several accounts /9/. De Michelis

Springer Series in Solid-State Sciences, Vol. 79 **Elementary Excitations in Quantum Fluids**
Editors: K. Ohbayashi · M. Watabe © Springer-Verlag Berlin, Heidelberg 1989

et al./10/ suggested that rotons get more and more populated as the temperature approaches toward T_λ and these, being single-particle-like excitations, may favor local order. This idea seems to be a fairly well accepted one.

It is then natural to ask: When rotons can spoil local order, why not the two-particle excitations do the same?

Attempts have been made earlier to understand the broad(multi excitation) peak observed in neutron scattering experiments by means of two-particle excitations. Wong /11/ observed that the multiphonon peak(when approximated by two-particle excitations) at high energy transfers is independent of statistics, depends only on the interparticle potential and has a long tail. This theory depends on two adjustible parameters. A later attempt by Tripathy et al. /12/ also has some mathematically unwelcome features. This situation outlines the necessity to study two-particle excitations in a more systematic manner.

3. The Structure Factor

Assuming that the density fluctuation operator acting on the ground state generates an approximate eigenstate of the Hamiltonian, Feynman obtained his celebrated form for the excitation energy which, for the first time, established a direct relationship of the excitations to the pair correlations. This did not connect the pair interactions with the excitation energy. Extension to higher temperatures was also not indicated. If one uses the well known Bogoliubov form for the excitation energy, the structure factor could be written as

$$S(k) = \{ 1 + 2 N V(k)/E_0(k) \}^{1/2} \tag{1}$$

where $E_0(k)$ is the free particle energy. Very few attempts have been made to calculate 'ab initio' the structure factor. One has to mention the approach due to Mihara and Puff /13/ which is based on the first few moments of the dynamic form factor $S(k,\omega)$ and leads to a non-linear integral equation for the structure factor. Sridhar and Vasudevan /14/ proposed an approach to evaluate $S(k)$ by taking into account the second branch observed in the spectrum also. The basic assumptions of this theory may be briefly outlined as follows:
  (i) the sum rules are saturated by single and two-particle excitations only
 (ii) the two-particle excitations are well defined and correspond to a rather sharp peak
(iii) the two-particle excitation energy is given by

$$\nu_2(k) = \begin{cases} a & \text{for } k \leq k_0 \\ E_0(k) & \text{for } k \geq k_0 \end{cases} \tag{2}$$

where a is a constant which may be chosen around 25K for $k_0 = 2.0A^{0-1}$.
(iv) The single particle excitation energy is assumed to be fairly well described by the Bogoliubov form.

The integral equation thus obtained yields a qualitatively satisfactory structure factor exhibiting all the necessary features in the weak coupling limit. Sridhar and Shanthi /15/ extended this approach to non-zero temperatures in which the single-particle energies and matrix elements were directly deduced from experiments. The main structure factor peak is found to be increasing as the temperature is increased from below T_λ towards T_λ. Broadening of the peak as the temperature is lowered is also obtained. The static density-density correlation function has a constant value of -1 for zero momentum transfer as expected from the compressibility sum rule. For momentum transfers of the order $2A^{o-1}$ this function decreases as the temperature is increased. This seems to confirm the prediction of de Michelis et al /10/ regarding the role of roton excitation and the proposal of hybridization of single and two-particle excitations due to Ruvalds and Zawadowski /16/.

4. Neutron Intensities

As has been mentioned earlier, the multiexcitation peak is broad and the integral equation obtained by Sridhar and Vasudevan has to be suitably modified /17/. The physical pair excitation state with total momentum k is supposed to be an admixture of states of relative momenta p with a weight f(k,p). Within RPA the two-excitation energy can be written as

$$\nu_2(k,p) = E_{Bog}(p_1) + E_{Bog}(p_2) \tag{3}$$

where the right side represents sum of Bogoliubov energies and p_1, p_2 are respectively given by (k/2) ± p. The present formulation does not invoke any specific ground state wave function and hence it is not possible to specify precisely the function f(k,p) by using a variational procedure. Thus we construct a model for this function using physical arguments. For large momentum values the single-particle components are uncorrelated and the weight function should give maximum contribution for small values of the relative momenta. Further this function is associated with the broadening of the peak and hence to life-time effects. Consequently the most probable choice is a Gaussian. This form is also suggested by the form of $S(k,\omega)$ for the thermal scattering of neutrons from an isotropic harmonic potential for small values of the wave vector. We thus choose

$$f(k,p) = A \; k^{3/2} \; \exp\left[-\alpha \, k^2 p^2/4\right] . \tag{4}$$

The parameter α is fixed from the condition

$$\int_0^\infty dk \; k^2 \; (S(k) - 1) = -2\pi^2\rho . \tag{5}$$

Sridhar and Shanthi /18/ have calculated the intensities at several momentum transfers by choosing a constant two-particle interaction which was chosen to reproduce the observed velocity of sound. It was also assumed that the two single particle components (that constitute the pair excitation) have their momenta either parallel or antiparallel to each other. The table below displays the calculated neutron intensities for several momentum transfers.

Table. Calculated peak position of neutron scattering intensities

Momentum transfers $/A^{o-1}/$	Peak position of the two-particle intensity /K/	
	Calculated	Experimental
0.5	18.0	19.35
0.6	21.0	20.07
0.8	25.0	24.50
1.0	30.0	26.83
1.13	21.9	22.70

A comparison with experiment shows that our values for $k = 0.5A^{o-1}$ agree well with experiment for almost all energy transfers. For $k = 0.6A^{o-1}$ the fit with smaller energy transfers is good while the tail part is slightly higher than the experiment. On the contrary for $k = 0.8A^{o-1}$ the tail part is slightly lower than the experimental values. The tail exactly agrees well with experiment for $k = 1.0A^{o-1}$. Though the peak position agrees well with the experiment, the qualitative fit to experimental data needs improvement for $k = 1.13A^{o-1}$.

5. More on Temperature Dependence of S(k)

We have also generalised our integral equation for the diffuse pair excitation peak to apply it to non-zero temperatures /19/. In this approach the temperature effects are introduced not only through the grand canonical averages but also through the renormalization of the effective interaction and the variation of the average kinetic energy per particle with temperature. The latter effects are best evaluated through the self energy of the temperature Greens functions. However, even for the case of a weakly interacting Bose system this is a complicated exercise /20/. One, therefore, resorts to an effective mass approximation which in general is momentum dependent. It turns out this momentum dependence is also rather involved and a constant effective mass appropriate to a given temperature, calculated from the velocity of sound at a fixed density, is used. The temperature dependence of the pair-excitation energy is introduced only through that of the single particle excitation energy by using eq(3).

Our numerical estimates bring out well the general qualitative dependence of S(k) on temperature. The quantitative fit is good for small ($k < 1.0^{o-1}$) as well as for large ($> 2.5A^{o-1}$) k values. The roton region requires a better quantitative fit. An interesting observation is that an improved RPA calculation for the pair excitation energy does not yield a significant variation in S(k). The fit may be improved by using a proper momentum dependent effective mass and by incorporating three particle correlations. The sum rule condition (5) is exactly satisfied by our S(k) whereas the experimental values observed by Robkoff et al /21/ do not satisfy (5). The static density-density correlation function displays the same qualitative behaviour as discussed in section 3.

The temperature has a significant role to play in determining pair correlations as brought out by the T dependence of the parameter α-a slight increase in α with T implying thereby a reduction in the range of k and p over which f(k,p) is significant.

6. Discussion

The theory outlined above brings out clearly the significant role played by two-particle excitations in determining S(k) and the static density-density response functions for small k and at low temperatures. More theoretical as well as experimental work is required on the two-particle excitations (along the lines indicated in ref.4) to get a better understanding. It is suspected that these excitations may also decide the anomalous dispersion of phonons at low momenta /22/.

An immediate requirement is to get a better and first principle determination of the weighting function f(k,p) introduced in eq(4). This is a rather difficult proposition since in our approach no explicit form of the ground state wave function is used.

A pressure dependent study will make a lot of information about the pair correlations and the effective pair interaction available. It is expected that f(k,p) may exhibit interesting features at higher pressures. A study along these lines is in progress and will be reported elsewhere.

<u>References</u>

1. R.A.Cowley and A.D.B.Woods: Can.J.Phys. <u>49</u>,177(1971).
2. A.Miller, D.Pines and P.Nozieres: Phys. Rev. <u>127</u>, 1452(1962)
3. T.J.Greytak and J.Yan: Phys. Rev. Lett. <u>22</u>,987(1969)
4. N.M.Blagoveshchenskii, E.B.Dokukin, Zh. A. Kozlov and V.A.Parfenov: JETP Lett. <u>30</u>,12(1979), ibid <u>31</u>,4(1980)
5. K.Ohbayashi and M.Udagawa: <u>Proc. Int. Conf. on Low Temp. Phys - LT 17,</u> ed by U.Eckern, A.Schmidt, <u>W.Weber</u> and H.Wuhl(North-Holland, Amsterdam 1984) p.1201
6. V.F.Sears and E.C.Svensson, Phys. Rev. Lett. <u>43</u>,2009(1979)
7. H.N.Robkoff, D.A.Ewen and R.B.Hallock: Phys. Rev. Lett. <u>43</u>,2006(1979)
8. G.J.Hylands, G.Rowlands and F.W.Cummings: Phys. Lett. <u>31A</u>,465(1970)
9. A.Griffin: Phys. Rev. <u>B22</u>,5193(1980); G.V.Chester and L.Reatto: Phys. Rev. <u>B22</u>, 5199(1980); A.L.Fetter: Phys. Rev. <u>B23</u>, 2425(1981); H.B.Ghassib and R.Sridhar: Phys. Lett.<u>100A</u>,198(1984)
10. C.de Michelis, G.L.Masserini and L.Reatto: Phys. Lett. <u>66A</u>,484(1978)
11. V.K.Wong: Phys. Lett. <u>61A</u>,454(1977)
12. D.N.Tripathy and M.Bhuyan: Phys. Lett. <u>86A</u>,303(1981)
13. N.Mihara and R.D.Puff: Phys. Rev. <u>174</u>,221(1968)
14. R.Sridhar and R.Vasudevan: Acta. Phys. Pol. <u>A61</u>, 81(1982)
15. R.Sridhar and A.Shanthi: Phys. Rev. <u>26B</u>,1252(1982)
16. J.Ruvalds and A.Zawadowski: Phys. Rev. Lett. <u>25</u>,333(1970) A.Zawadowski, J.Ruvalds and J.Solana: Phys. Rev. <u>A5</u>,399(1972)
17. R.Sridhar: Acta. Phys. Pol. <u>A65</u>,9(1984)
18. R.Sridhar and A.Shanthi: Phys. Lett. <u>93A</u>,198(1983)
19. R.Sridhar and A.Shanthi: Physica <u>137</u>,337(1986)
20. A.Isihara: Physica <u>106B</u>, 161(1981)
21. H.N.Robkoff and R.B.Hallock: Phys. Rev. <u>B25</u>,1572(1982)
22. R.Sridhar: Phys. Rep. <u>146</u>,260(1987)

Two-Component Character of $S(k, \omega)$ in Superfluid ^{4}He

K. Ishikawa[1] *and K. Yamada*[2]

[1]Department of Physics, Yokohama City University, Kanazawa-ku,
 Yokohama 236, Japan
[2]College of General Education, Nagoya University, Nagoya 464, Japan

WOODS and SVENSSON[1] analyzed the experimental results of
the dynamical sturucture factor, S(k,ω) for superfluid ^{4}He at
finite temperatures and proposed in 1978 that S(k,ω) is
described as the sum of two components:

$$S(k,\omega) = \bar{\rho}_s S_s(k,\omega) + \bar{\rho}_n S_n(k,\omega), \qquad (1)$$

where $\bar{\rho}_s = \rho_s/\rho$ and $\bar{\rho}_n = \rho_n/\rho$, and ρ , ρ_s and ρ_n denote total,
superfluid and normal fluid density, respectively.   Later
one of us (K.YAMADA[2]) investigated the response functions in
superfluid ^{4}He on the basis of the two fluid dynamics and
showed that the S(k,ω), in the small but collisionless region
of wave vector, can be described as the sum of two components,
supporting the analysis proposed by Woods and Svensson.
However the above treatment is restricted in temperatures
lower than 1K and cannot be compared with the experimental
results, because the contributions of phonons are taken into
account as thermal excitations but those of rotons are not
considered.

Here we try to extend the preceding work in the form
applicable in temperatures higher than 1K, taking into account
rotons as thermal excitations, and examine whether the Woods
and Svensson's proposal holds or not theoretically.   This
problem has been analyzed before by TALBOT and GRIFFIN [3].
Their work was based on the Green function method in finite
temperature and showed negative answer to the two-component
structure of S(k,ω).   Our theory is based on the two fluid
dynamics is simpler than that of Talbot and Griffin.

We use three basic equations of two fluid dynamics[4]:

Conservation law of a mass density with a mass current
$\mathbf{j} = \rho \mathbf{v}_s + \Sigma \mathbf{p} N(p)$, equation of motion for the superfluid velocity

---

Springer Series in Solid-State Sciences, Vol. 79   **Elementary Excitations in Quantum Fluids**   155
Editors: K. Ohbayashi · M. Watabe   © Springer-Verlag Berlin, Heidelberg 1989

$\mathbf{v}_S$ and the kinetic equation for the roton distribution function N(p). To obtain the response function from the two fluid dynamical equations, they are regarded as operator equations for ρ, $\mathbf{v}_S$ and N(p).

Using the linearized and Fourier-transformed equations, the equations of motion for the response function defined by

$$\chi_{AB}(k,t) = -i\theta(t)\langle[A_k(t),B_{-k}(t)]\rangle,\ \text{are written as}$$

$$-i\omega\chi_{\rho\rho}(k,\omega) - \rho k^2\chi_{\phi\rho}(k,\omega) + i\sum_p \Sigma(\mathbf{kp})\chi_{N\rho}(p;k,\omega) = 0, \qquad (2)$$

$$-i\omega\chi_{\phi\rho}(k,\omega) + (\tilde{c}^2/\rho)\chi_{\rho\rho}(k,\omega) + (cp_0/\rho)\sum_p U_p\chi_{N\rho}(p;k,\omega) = 1, \qquad (3)$$

$$-i(\omega+i\Gamma_p-\mathbf{kv}_p)\chi_{N\rho} - i(\mathbf{kv}_p)N_p'[(cp_0/\rho)U_p\chi_{\rho\rho}+i(\mathbf{kp})\chi_{\phi\rho}] = 0. \qquad (4)$$

Here, ϕ_k is the velocity potential defined by $\mathbf{v}_{sk}=i\mathbf{k}\phi_k$.

$N_p'= dN^0(p)/dE_p$ is the derivative of the roton distribution function at equilibrium $N^0(p)$ by the roton energy $E_p=\Delta +(p-p_0)^2/2\mu$.

$\mathbf{v}_p = (\mathbf{p}-\mathbf{p}_0)/\mu$ is the roton velocity. $(cp_0/\rho)U_p$ is phonon-roton coupling given by

$$(cp_0/\rho)U_p = \partial E_p/\partial\rho = \partial\Delta /\partial\rho -((p-p_0)/\mu)\partial p_0/\partial\rho$$

$$-((p-p_0)^2/(2\mu^2))\partial\mu /\partial\rho, \qquad (5)$$

$$\tilde{c}^2 =c^2 + \sum\partial^2 E_p/\partial^2 N^0(p).$$

Γ_p is the inverse roton life time. In order that these equations are valid, the wave vector k is restricted in the collisionless phonon region. Solving the equations of motion for the response function, we find

$$\chi_{\rho\rho}(k,\omega) = -Q_1(k,\omega)/D(k,\omega) \qquad (6)$$

where

$$D(k,\omega) = P^2(k,\omega) - Q_1(k,\omega)Q_2(k,\omega), \tag{7}$$

$$Q_1(k,\omega) = \rho k^2 - \sum_p (\mathbf{kp})^2 (\mathbf{kv}_p) N_p'(\omega + i\Gamma_p - \mathbf{kv}_p)^{-1}$$

$$= \rho k^2 [\bar{\rho}_s - (1/2\pi^2)\int dp\, p^4 N_p'(\omega/kv_p)^2 \xi(k\omega, v_p)], \tag{8}$$

$$Q_2(k,\omega) = (c^2/\rho)[(\tilde{c}/c)^2 + (1/2\pi^2\rho)\int dp\, p^2 p_0^2 N_p' U_p^2 \xi(k\omega, v_p)], \tag{9}$$

$$P(k,\omega) = \omega + (ck/2\pi^2\rho)\int dp\, p^4 U_p N_p'(\omega/kv_p)\xi(k\omega, v_p). \tag{10}$$

To evaluate the above integrals, we use the approximation to replace v_p by thermal velocity of roton $v_T = (k_B T/\mu)^{1/2}$.

Using the well known relations

$$\rho_n = -\sum_p (\mathbf{kp})^2 N_p'/k^2 \quad \text{and} \quad \rho - \rho_n = \rho_s,$$

we obtain the density-density response function as

$$\chi_{\rho\rho} = (-k^2\rho_s + 3\rho_n(\omega/v_T)^2\xi)/D(k,\omega). \tag{11}$$

Here, $\xi(k\omega, v_T) = (\omega/2kv_T)[\log(\omega + i\Gamma_p + kv_T) - \log(\omega + i\Gamma_p - kv_T)]. \tag{12}$

The numerator of the density-density response function $\chi_{\rho\rho}$ is the sum of a superfluid component and a normal fluid component and $\chi_{\rho\rho}$ is easily separated like eq.(1).

The dynamical structure factor $S(k,\omega)$ is proportional to the imaginary part of $\chi_{\rho\rho}$ and has two-component structure as

$$S_s(k,\omega) = \frac{\rho k^2 2\omega\Gamma}{(1 - e^{-\beta\omega})[(\omega^2 - \varepsilon_k^2)^2 + (2\omega\Gamma)^2]} \tag{13}$$

$$S_n(k,\omega) = \frac{3\rho(\omega/v_T)^2[(\omega^2 - \varepsilon_k^2)\xi_I + 2\omega\Gamma\xi_R]}{(1 - e^{-\beta\omega})[(\omega^2 - \varepsilon_k^2)^2 + (2\omega\Gamma)^2]} \tag{14}$$

where

$$\varepsilon_k^2 = (ck)^2 [\bar{\rho}_s (\tilde{c}/c)^2 + 3\bar{\rho}_n \beta \xi_R + 9\bar{\rho}_n^2 \alpha (\xi_R^2 - \xi_I^2)] \tag{15}$$

$$2\omega\Gamma = 3\bar{\rho}_n (ck)^2 \xi_I (\beta + 6\bar{\rho}_n \alpha\xi_R), \tag{16}$$

$$\beta = \bar{\rho}_s [A^2 + (B^2 - 2AG)v_T^2 + G^2 v_T^2] + (\omega/ck)^2 [(\tilde{c}/v_T)^2 - 2cB] \tag{17}$$

$$\alpha = (\omega/kv_T)^2 (A^2 - 2AGv_T^2 + G^2 v_T^4). \tag{18}$$

Here, ξ_R and ξ_I are the real and imaginary part of ξ, and A, B and G are defined by $A = (\rho/p_0 c)\partial\Delta/\partial\rho$, $B = (\rho/p_0 c)\partial p_0/\partial\rho$ and $G = (\rho/2p_0 c)\partial\mu/\partial\rho$. The $S_s(k,\omega)$ has a resonance at ε_k with a width $2\omega\Gamma$. The numerator of $S_n(k,\omega)$ is complex but if the temperature above T_λ is considered and $\bar{\rho}_s = 0$ and $\bar{\rho}_n = 1$ are put into (14), the $S_n(k,\omega)$ becomes a structure similar to Fermi liquid as

$$S_n(k,\omega) = 3\rho\xi_I [1 + g_2(\xi_R^2 + \xi_I^2)]/[v_T^2(1 - e^{-\beta\omega})]D(k,\omega) \tag{19}$$

with

$$D(k,\omega) = [1 - g_1\xi_R - g_2(\xi_R^2 - \xi_I^2)]^2 + \xi_I^2(g_1 + 2g_2\xi_R)^2, \tag{20}$$

which indicates a zero-sound-like resonance of roton-phonon gas. However, as easily seen from (14), below T_λ, the $S_n(k,\omega)$ has a same resonance at ε_k with a width proportional to Γ as $S_s(k,\omega)$ and a different structure from (19) which is characteristic of normal fluids.

Therefore the simple separation of the dynamical structure factor $S(k,\omega)$ into the superfluid and normal parts like (13) and (14) does not correspond to the experimental separation proposed by Woods and Svensson. Theoretical support for two-component character of the dynamical structure factor is not yet given in this work. But there are some possibilities of separation of the $S(k,\omega)$ into the superfluid and normal parts different from (13) and (14). To conclude whether the Woods and Svensson proposal is correct or not, we must investigate the other separations further. The details will be given elsewhere.

1. A.D.B.Woods, E.C.Svensson: Phys. Rev. Letters, $\underline{41}$, 974
   (1978)
2. K.Yamada: Prog. Theor. Phys. $\underline{63}$, 715 (1980)
3. E.Talbot, A.Griffin: Phys. Rev. $\underline{B29}$, 2531(1984)
4. L.M.Khalatnikov: <u>An Introduction to the Theory of
   Superfluidity</u>,(W.A.Benjamin Inc, New York,1965)

Theory of Temperature Dependence of Elementary Excitation Energy

S. Sasaki

Department of Physics, College of General Education, Osaka University, Toyonaka, Osaka 560, Japan

Assuming that the total Hamiltonian of liquid helium is completely diagonalized by a unitary operator, we can introduce "dressed-boson" operators as the unitary transformation of creation and annihilation operators of a helium atom. Then we can make a new viewpoint for the elementary excitation of liquid helium on the basis of the concept of the "dressed-boson". The temperature dependence of the elementary excitation energy is studied. It is verified that the elementary excitation with a small momentum softens as the temperature approaches λ-point. The detection of the softening is proposed.

1. INTRODUCTION

Hitherto, there have been many developments [1-5] in the investigation of the elementary excitation of liquid HeII. The elementary excitation is regarded as the production of the Landau quasiparticle. In this paper, we propose a new viewpoint for the elementary excitation.

If we could transform the total Hamiltonian of liquid helium completely into a diagonal form, we would be naturally led to this viewpoint.

(Assumption 1) Let us first assume that the total Hamiltonian is completely diagonalized by a unitary operator U , although we do not need the explicit form of U. It may be considered that this assumption is quite reasonable but clarifies nothing. However, if we add one more assumption, we can make a new viewpoint for the elementary excitation.

(Assumption 2) The ground state of liquid helium is $U(a_o^*)^N|0\rangle$ where a_o^* is a creation operator of one helium atom with zero momentum, N is the total number of helium atoms, and $|0\rangle$ denotes a vacuum state.

From the first assumption, we can introduce new creation and annihilation operators A_p^*, A_p which are defined as

$$A_p^*=Ua_p^*U^*, \quad A_p=Ua_pU^*,\tag{1}$$

where a_p^* and a_p are creation and annihilation operators of one helium atom with momentum p , and satisfy the following commutation relations:

$$[a_p, a_q^*]=\delta_{p,q}, \quad [a_p, a_q]=[a_p^*, a_q^*]=0.\tag{2}$$

Let us call these new operators A_p^* and A_p the creation and annihilation operators of a "dressed-boson", respectively. Then,

Springer Series in Solid-State Sciences, Vol. 79 **Elementary Excitations in Quantum Fluids**
Editors: K. Ohbayashi · M. Watabe © Springer-Verlag Berlin, Heidelberg 1989

we can express the ground state in terms of the dressed-boson operators, as follows:

$$|\text{Ground State}> = \frac{1}{\sqrt{N!}} (A_0^*)^N |0>, \tag{3}$$

where we use the unitary property of U and $U|0> = |0>$ because the vacuum state is an eigen-state of the total Hamiltonian. Similarly, a single excitation state is expressed as

$$|\text{Single Excitation}> = \frac{1}{\sqrt{(N-1)!}} A_p^* (A_0^*)^{N-1} |0>. \tag{4}$$

Namely, the elementary excitation means that one "dressed-boson" with zero momentum is annihilated and one "dressed-boson" with momentum p is created. This viewpoint for the elementary excitation and the concept of the "dressed-boson" give a new explanation for the phenomena of liquid helium. In the previous paper[6], we derived many characteristic properties of liquid helium from this viewpoint. In this paper, we clarify the temperature dependence of excitation.

2. GENERAL FORM OF THE TOTAL ENERGY

Let us write down the total Hamiltonian of many helium atoms,

$$H = \sum_p \frac{p^2}{2m} a_p^* a_p + \frac{1}{V} \sum_{p,q,k} g(k) a_{p+k}^* a_{q-k}^* a_p a_q, \tag{5}$$

where m and $g(k)$ are mass and interatomic potential of helium atoms respectively, and momenta p, q and k satisfy the periodic boundary condition as $p_x = (2\pi\hbar/L) \times \text{integer}$, $p_y = (2\pi\hbar/L) \times$ integer, and $p_z = (2\pi\hbar/L) \times$ integer in the volume $V = L^3$.

From the assumption 1, all eigen-states of H and their eigen-equations are given by

$$H A_{q_1}^* \cdots A_{q_N}^* |0> = E_{q_1 \cdots q_N} A_{q_1}^* \cdots A_{q_N}^* |0> \tag{6}$$

where $E_{q_1 \cdots q_N}$ denotes a total energy, and we use the fact that there is no bound state of helium atoms. This eigen-equation illustrates that the total energy can be expressed in terms of the dressed-boson number n_q, which is defined by

$$n_q = A_q^* A_q = U a_q^* a_q U^*. \tag{7}$$

Since the Hamiltonian (5) is covariant under the Galilean transformations, the total energy is also Galilean-covariant. Therefore, the total energy is expressed as

$$E = \sum_p \frac{p^2}{2m} n_p + \frac{1}{V} \sum_{p,q} h_2(p-q) n_p n_q$$
$$+ \frac{1}{V^2} \sum_{p,q,r} h_3(p-q, p-r) n_p n_q n_r + \cdots + E_G(N, V), \tag{8}$$

where functions h_l are only dependent upon the momentum differences (which are invariant under the Galilean transformations), E_G is the ground state energy, and $h_2(0)=h_3(0,0)=\cdots=0$ because of assumption 2.

Now, let us rewrite expression (8) in terms of new functions f_ℓ , which are linear combinations of $\{h_i\}$ and have such properties as

$$f_\ell(p_1-p_2, p_1-p_3, \cdots\cdots, p_1-p_\ell; \widetilde{\rho})=0 \text{ for } p_i=p_j, \tag{9}$$

where $\widetilde{\rho}=N/V$. Namely the value of f_ℓ vanishes when two momenta p_i and p_j take the same value. For example, the function f_2 is given by

$$f_2(p-q;\widetilde{\rho})=2\,h_2(p-q)+\frac{N}{V}\,[h_3(p-q,p-q)+$$
$$+h_3(p-q,0)+h_3(0,p-q)]+\cdots\cdots . \tag{10}$$

These re-expressions of h_ℓ give the following representation of the total energy:

$$E=\sum_p \frac{p^2}{2m}\,n_p+\frac{1}{2V}\sum_{p,q}f_2(p-q;\widetilde{\rho})\,n_p n_q$$
$$+\frac{1}{3V^2}\sum_{p,q,r}f_3(p-q,p-r;\widetilde{\rho})\,n_p n_q n_r+\cdots\cdots+E_G(N,V). \tag{11}$$

This re-expression of the total energy is verified in detail in a coming full paper[7].

Now, we show that these unknown functions f_ℓ are determined from the excitation energy at absolute zero temperature.

First, we write the number distributions of the dressed-bosons before and after excitation at zero temperature:
(Before excitation)

$$n_q=0 \text{ for } q\neq0, \quad n_0=N \text{ for } q=0. \tag{12}$$

(After excitation)

$$n'_q=0 \text{ for } q\neq0 \text{ and } q\neq p,$$
$$n'_p=1 \text{ for } q=p, \tag{13}$$
$$n'_0=N-1 \text{ for } q=0.$$

Second, the elementary excitation energy ε_p^0 at zero temperature is given by

$$\varepsilon_p^0=E(\{n'_q\})-E(\{n_q\})=\frac{p^2}{2m}+\frac{N-1}{V}f_2(p;\widetilde{\rho}),$$

where we use Eqs.(9 and 11) and rotational invariance of f_2 . In the thermodynamic limit ($V\to\infty$ in fixing the value of $\widetilde{\rho}=N/V$), the above equation leads to the following relation beween f_2 and ε_p^0 :

$$f_2(p;\widetilde{\rho})=\widetilde{\rho}^{-1}(\varepsilon_p^0-\frac{p^2}{2m}). \tag{14}$$

The other functions $f_\ell(\ell\geq3)$ can be determined from the experimental data for the interaction energies among multi-excitations.

162

3. TEMPERATURE DEPENDENCE OF EXCITATION ENERGY

In the previous paper [6], we derived an integral equation which determines the number distribution $\{\bar{n}_p\}$ of the dressed-bosons at an equilibrium state. Moreover, we obtained the solutions of the number distribution, which has a Bose-condensate at lower temperature than a transition temperature T_λ, but has no condensate at higher temperature than T_λ. Here, we represent the number distribution by $\bar{n}_p(T,P)$ in the case for zero normal-velocity and zero super-velocity (here, T and P denote temperature and pressure of liquid helium). Then, before excitation, the number distribution is $\bar{n}_p(T,P)$, and, after excitation, the number distribution becomes

$$n'_q = \bar{n}_q(T,P) \text{ for } q \neq 0 \text{ and } q \neq p,$$
$$n'_p = \bar{n}_p(T,P)+1 \text{ for } q=p,$$
$$n'_o = \bar{n}_o(T,P)-1 \text{ for } q=0. \tag{15}$$

Accordingly, the excitation energy $\varepsilon_p(T,P)$ at a temperature T and a pressure P is given by

$$\varepsilon_p(T,P) = E(\{n'_q\}) - E(\{\bar{n}_q(T,P)\}). \tag{16}$$

Substituting Eq.(11) into the right-hand-side of Eq.(16), and using the rotational invariance of f_2, we obtain

$$\varepsilon_p(T,P) = \frac{p^2}{2m} + \widetilde{\rho}_s f_2(p\,;\widetilde{\rho}) + \frac{1}{V}\sum_{q\neq 0}\left[f_2(p-q\,;\widetilde{\rho})-f_2(-q\,;\widetilde{\rho})\right]\bar{n}_q(T,P)$$
$$+ \frac{1}{3V^2}\sum_{q,r}\left[f_3(p-q,p-r)-f_3(-q,-r)+f_3(q-p,q-r)\right.$$
$$\left.-f_3(q,q-r)+f_3(q-r,q-p)-f_3(q-r,q)\right]\bar{n}_q\bar{n}_r+\cdots\cdots, \tag{17}$$

where $\widetilde{\rho}_s = \bar{n}_o(T,P)/V$ denotes a number density of the condensed dressed-bosons and takes a positive value at $T<T_\lambda$ but zero value at $T\geq T_\lambda$ in the thermodynamic limit. This expression (17) shows that the elementary excitation energy varies with a change of temperature.

4. ELEMENTARY PHONON VELOCITY

From the elementary excitation energy (17), we can calculate the phonon velocity $C(T,P)$ at temperature T and pressure P :

$$C(T,P) \equiv \lim_{p\to 0}\varepsilon_p(T,P)/|p| = \widetilde{\rho}_s\lim_{p\to 0}|p|^{-1}f_2(p\,;\widetilde{\rho})+V^{-1}\lim_{p\to 0}|p|^{-1}\times$$
$$\times\sum_{q\neq 0}\left[f_2(p-q\,;\widetilde{\rho})-f_2(-q\,;\widetilde{\rho})\right]\bar{n}_q(T,P)+\widetilde{\rho}_s\lim_{p\to 0}(3V|p|)^{-1}\left\{\sum_r\left[f_3(p,p-r)\right.\right.$$
$$\left.+f_3(-p,-r)+f_3(-r,-p)\right]\bar{n}_r+\sum_q\left[f_3(p-q,p)+f_3(q-p,q)+f_3(q,q-p)\right]\bar{n}_q\bigg\}$$
$$+\lim_{p\to 0}(3V^2|p|)^{-1}\sum_{q\neq 0,r\neq 0}\left[f_3(p-q,p-r)-f_3(-q,-r)+f_3(q-p,q-r)\right.$$
$$\left.-f_3(q,q-r)+f_3(q-r,q-p)-f_3(q-r,q)\right]\bar{n}_q\bar{n}_r+\cdots\cdots, \tag{18}$$

where we use the property (9) of the function f_i. In this equation, the second and final terms in the right-hand side become zero because of the symmetric property of the number distribution $\bar{n}_q$ for the momentum q . For example

$$\lim_{p\to0}(V|p|)^{-1}\sum_{q\neq0}[f_2(p-q)-f_2(-q)]\bar{n}_q$$

$$=\lim_{p\to0}(V|p|)^{-1}\sum_{q\neq0}[-p\cdot\mathrm{grad}_q\,f(-q)]\bar{n}_q=0$$

where we use the property $\bar{n}_q=\bar{n}-q$. Thus, the phonon velocity $C(T,P)$ is proportional to the number density $\tilde{\rho}_s$ of the superfluid (note: $\tilde{\rho}_s$ is different from the mass density of superfluid), as follows:

$$C(T,P)=\tilde{\rho}_s\Big\{C_0(\tilde{\rho})/\tilde{\rho}+\lim_{p\to0}(3V|p|)^{-1}\sum_q[f_3(p,p-q)+f_3(p-q,p)$$

$$+f_3(-p,-q)+f_3(-q,-p)+f_3(q-p,q)+f_3(q,q-p)]\bar{n}_q(T,P)+\cdots\cdots\Big\},\qquad(19)$$

where $C_0(\tilde{\rho})$ denotes a phonon velocity at zero temperature namely,

$$C_0(\tilde{\rho})=\lim_{p\to0}\varepsilon_p^0/|p|=\lim_{p\to0}\tilde{\rho}f_2(p;\tilde{\rho})/|p|\quad.\qquad(20)$$

Equation (19) indicates that the phonon velocity $C(T,P)$ vanishes at the λ-point (see Fig.1). Namely, $C(T,P)$ goes away from the experimental values of the first sound velocity, as the temperature approaches the λ-point. Let us call this elementary excitation with a small momentum by "elementary phonon", the velocity of which is $C(T,P)$.

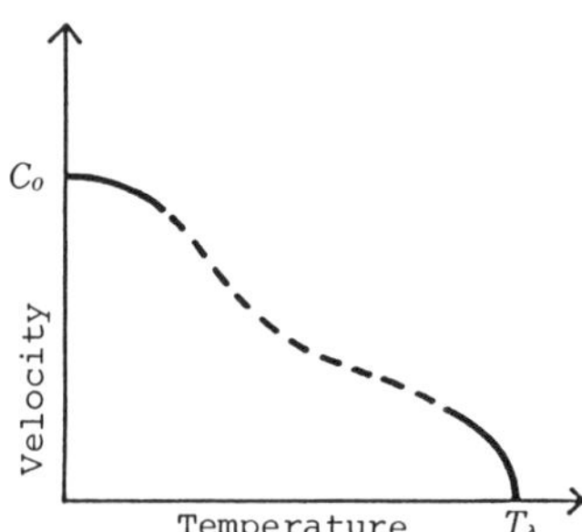

Fig. 1 Typical behaviour of
elementary phonon velocity

Now, the temperature dependence of the "elementary phonon velocity" is derived from a non-linear form of the total energy with respect to the dressed-boson number. Namely, the second and following terms in the total energy (11) are non-linear with respect to the number of the dressed-bosons because of Galilean-invariance, and, therefore, the elementary phonon velocity necessarily depends upon the number distribution of the other dressed-bosons. Consequently the "elementary phonon velocity" varies with a change of temperature.

Let us discuss the temperature dependence of the "elementary phonon velocity" near the λ-point. In the forthcoming paper, we discuss the detailed behaviour of $\tilde{\rho}_s$ near T_λ , which results in

$$\widetilde{\rho}_s \propto (T_\lambda - T)^{1/3} \quad \text{for} \quad T_\lambda - T \ll T_\lambda . \tag{21}$$

(note: this behaviour of $\widetilde{\rho}_s$ is different from the temperature dependence of the mass density of superfluid) Therefore, the "elementary phonon velocity" becomes

$$C(T,P) \xrightarrow[T \to T_\lambda]{} (\text{constant}) \times (T_\lambda - T)^{1/3}. \tag{22}$$

Thus, we know the behaviour of $C(T,P)$ near the λ-point.

Here, it should be noted that this "elementary phonon" necessarily exists, if the assumptions mentioned above hold in liquid helium. Therefore, reexamining the data of various experiments, we find out the interesting experiments, namely Brillouin scattering, among them. In the results of the experiments [8], the linewidthes of the second sound are too small in comparison with the values of the hydrodynamic calculations in the temperature range $T_\lambda - T < 10$ mdeg. Moreover, the temperature dependence of the second sound velocity is in accord with that of the "elementary phonon velocity", as seen in Eq.(22).

Therefore, one possible answer to the question "what the elementary phonon should be" is as follows: The observations of the second sound in the Brillouin scatterings were the detection of the "elementary phonon". The other possible answer is that another new phonon should be discovered in further experiments.

Next, we discuss the dispersion curve of the elementary excitation. Since the softening of the "elementary phonon" occurs as the temperature approaches the λ-point, the dispersion curve of the elementary excitation varies with temperature as in Fig. 2. If the instrumental linewidth of neutron scattering will become smaller, and the measurements will be carried out in the smaller momentum transfer (about 10^{-3} Å^{-1}), this softening of the "elementary phonon" might be detected. I hope some experimentalist would take interest in such experiments.

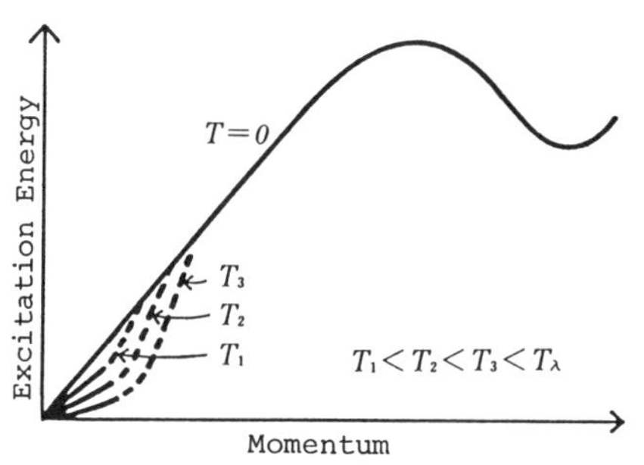

Fig. 2 Temperature dependence of elementary excitation
Enlarged near zero momentum

REFERENCES

1. L.D. Landau: J. Phys. USSR **5**, 71 (1941)
2. R.P. Feynman: Phys. Rev. **94**, 262 (1954)
3. A.D.B. Woods and R.A. Cowley: Rep. Prog. Phys. **36**, 1135 (1973)

4. H.W. Jackson: Phys. Rev. $\underline{B19}$, 2556 (1979)
5. E.C. Svensson, V.F. Sears and A. Griffin: Phys. Rev. B$\underline{23}$, 4493 (1981)
6. S. Sasaki: Sci. Rep., Col. Gen. Educ. Osaka Univ. $\underline{35}$-2,1 (1986)
    S. Sasaki: JJAP Supplement, Proceedings of LT-18, $\underline{26-3}$, 23 (1987)
7. S. Sasaki: in preparation
8. G. Winterling, F.S. Holmes and T.J. Greytack: Phys. Rev. Lett. $\underline{30}$, 427 (1973)
    G. Winterling, J. Miller and T.J. Greytak: Phys. Lett. $\underline{48A}$, 343 (1974)

Eulerian and Lagrangian Phonons

A. Thellung

Institut für Theoretische Physik der Universität Zürich, Schönberggasse 9,
CH-8001 Zürich, Switzerland

Hydrodynamic motion can be described either in the formulation of
Euler or of Lagrange. Quantization leads to phonons, which in
Euler's case carry a momentum $\hbar\underline{k}$ ($\underline{k}$ = wave vector) and in Lagrange's case have no momentum. In spite of this difference,
physically observable effects must be the same since the two
representations are equivalent. This conclusion is confirmed in
an explicit comparison of the two cases for the microscopic
derivation of the two-fluid equations in liquid He II.

1. INTRODUCTION

The differences between the formulations of hydrodynamics by
Euler (in local coordinates) and by Lagrange (in material coordinates) (see e.g. [1]) are briefly laid out in Section 2. In
Section 3 quantization is carried out, leading to the concept of
phonons. An Eulerian phonon of wave vector $\underline{k}$ turns out to have
a momentum $\hbar\underline{k}$, a Lagrangian phonon carries no momentum.

This fact gives rise to the question whether Euler's and Lagrange's formulations lead to different physical results. We mention two examples: (i) For the microscopic derivation of two-
fluid equations it is usually considered essential that an ele-
mentary excitation of wave vector $\underline{k}$ carries a momentum $\hbar\underline{k}$ [2]-[4].
(ii) Another example is the sound pressure if it is calculated
in analogy to the light pressure. Since the latter can be obtained as the result of the photons transferring their momenta to an
absorbing or reflecting screen it is hard to see how momentumless
phonons can produce a sound pressure. Therefore, intuitively one
is inclined to think that the two kinds of phonons lead to different physical results.

On the other hand, Euler's and Lagrange's formulations of hydrodynamics are equivalent. In fact, VAN SAARLOOS [5] has found
a canonical transformation relating the two representations. As
a consequence, for the same initial and boundary conditions physically observable effects must be the same in both formulations.

In spite of this clear conclusion one feels somewhat puzzled,
and it is most desirable to see in explicit detail how this result arises in the above-mentioned examples. For the two-fluid
equations this is shown in Section 4. For the sound pressure
the proof is given in [6].

---

2. LAGRANGIAN AND EULERIAN COORDINATES

A <u>Lagrangian</u> (material) coordinate $\underline{a}$ labels a given element of fluid (independent of its present position); it equals e.g. its equilibrium position when the fluid is at rest. An <u>Eulerian</u> (local) coordinate $\underline{x}$ characterizes a fixed point in space (independent of which element of fluid passes through that point). The connection between $\underline{x}$ and $\underline{a}$ is made by saying that $\underline{x}$ is the position at time t of the element of fluid whose equilibrium position is $\underline{a}$. Therefore the relation between $\underline{x}$ and $\underline{a}$ is

$$\underline{x} = \underline{a} + \underline{\xi}\,(\underline{a},t) \; , \tag{2.1}$$

where $\underline{\xi}\,(\underline{a},t)$ is the displacement vector at time t of the element of fluid labeled by $\underline{a}$. In Lagrange's formulation, $\underline{a}$ and t are taken as the independent variables; in Euler's formulation, the independent variables are $\underline{x}$ and t . Equ.(2.1) gives $\underline{x}$ as a function of $\underline{a}$ and t ; we assume that the inverse function $\underline{a}(\underline{x},t)$ exists. The velocity field in Lagrange's formulation is

$$\underline{v}_L(\underline{a},t) = \partial\,\underline{x}(\underline{a},t)/\partial t = \partial\,\underline{\xi}\,(\underline{a},t)/\partial t \; , \tag{2.2}$$

whereas Euler's velocity field $\underline{v}_E$ at $\underline{x},t$ is defined as

$$\underline{v}_L(\underline{a}(\underline{x},t),t) =: \underline{v}_E(\underline{x},t) \; . \tag{2.3}$$

In <u>Euler's</u> formulation the dependent variables are the velocity field $\underline{v}(\underline{x},t)$ and the density $\varrho\,(\underline{x},t)$; the pressure p is assumed to be a given function of ϱ (adiabatic law). The basic equations of motion [1] are Euler's equation

$$\partial\,\underline{v}/\partial t + (\underline{v}\cdot\underline{\nabla})\underline{v} = - (1/\varrho)\,\underline{\nabla}\,p \tag{2.4}$$

and the equation of continuity

$$\partial\varrho/\partial t + \underline{\nabla}\cdot(\varrho\,\underline{v}) = 0 \; . \tag{2.5}$$

In <u>Lagrange's</u> formulation the dependent variables are $\underline{x}(\underline{a},t)$, the position of the element of fluid $(\underline{a})$ at time t , and the density $\varrho\,(\underline{a},t)$, the pressure being again a function of ϱ . The basic differential equations [1] are Lagrange's equation of motion and an equation for the density, i.e.

$$(\partial^2 x_i/\partial t^2)(\partial x_i/\partial a_k) = - (1/\varrho)\,\partial p/\partial a_k \; , \tag{2.6}$$

$$\varrho_0/\varrho = \|\,\partial x_i/\partial a_k\| \; , \tag{2.7}$$

where $\|\partial x_i / \partial a_k\|$ means the Jacobian and ρ_0 the equilibrium density. (2.7) follows from the equation of conservation of mass,

$$\rho \, d^3x = \rho_0 \, d^3a \; . \tag{2.8}$$

In fluid mechanics one usually prefers Euler's form; in solid state physics one normally works with Lagrangian coordinates.

3. QUANTIZATION

3.1. Eulerian Phonons

Since we are only interested in (longitudinal) phonons we can restrict ourselves to vortex-free velocity fields, which we write as

$$\underline{v}_E = - \underline{\nabla}\varphi + \underline{U}_E \tag{3.1}$$

(φ = velocity potential, $\underline{U}_E$ = uniform translation). Euler's equations of motion can be obtained from a canonical formalism [7]. The Hamiltonian is

$$H_E = \int_V d^3x \; [\tfrac{1}{2} \, \underline{v}_E \, \rho \, \underline{v}_E + \frac{c^2}{2\rho_0} \, (\rho - \rho_0)^2 + \ldots \,] \; , \tag{3.2}$$

where the potential energy density is expanded in powers of $\rho - \rho_0$, the deviation of the density ρ from its (constant) equilibrium value ρ_0 . c is the speed of sound. ρ and φ are canonically conjugate fields. <u>Quantization</u> is straightforward [7]. Assuming periodic boundary conditions in a cube of volume V , we use a spatial Fourier decomposition [terms $\propto \exp(i\underline{k}\,\underline{x})$] of ρ and φ to diagonalize the harmonic part of H_E ,

$$H_E = \tfrac{1}{2} \rho_0 V \underline{U}_E^2 + \sum_{\underline{k}} n_{\underline{k}} \, (\varepsilon_{\underline{k}} + \hbar\underline{k}\,\underline{U}_E) + \text{anharm. terms} \; , \tag{3.3}$$

where $\varepsilon_{\underline{k}} = \hbar w_{\underline{k}} = \hbar c\,|\underline{k}|$, and $n_{\underline{k}} = 0,1,2,\ldots$ is the number of phonons of wave vector $\underline{k}$. The total momentum is

$$\underline{P}_E = \int_V d^3x \, \rho \, \underline{v}_E = \rho_0 V \underline{U}_E + \sum_{\underline{k}} n_{\underline{k}} \, \hbar\underline{k} \; . \tag{3.4}$$

Later, we shall need the spatially averaged velocity

$$\underline{u}_{sE} = \frac{1}{V} \int_V d^3x \, \underline{v}_E = \underline{U}_E \; . \tag{3.5}$$

(3.4) and (3.5) are exact. They show that <u>Eulerian phonons have momenta $\hbar\underline{k}$, but give no contribution to the average velocity.</u>

3.2. Lagrangian Phonons

Since the linearized versions of Euler's and Lagrange's equations are the same, the harmonic part of the Hamiltonian must be the same. Thus we have in Lagrange's case

$$\underline{v}_L = - \underline{\nabla}_a \phi + \underline{U}_L \; , \tag{3.6}$$

$$H_L = \int_V d^3a \; [\tfrac{1}{2} \rho_0 \, \underline{v}_L^2 + \frac{c^2}{2\rho_0} (\rho - \rho_0)^2$$
$$+ \underline{\text{different}} \text{ anharmonic terms}]. \tag{3.7}$$

ρ and ϕ are canonically conjugate. Spatial Fourier decomposition [terms $\propto \exp(i\underline{k}\,\underline{a})$!] and quantization yield for the Hamiltonian, the total momentum, and for the average velocity:

$$H_L = \tfrac{1}{2} \, \rho_0 V \underline{U}_L^2 + \sum_{\underline{k}} n_{\underline{k}} \, \varepsilon_{\underline{k}} + \text{ anharmonic terms}, \tag{3.8}$$

$$\underline{P}_L = \int_V d^3a \; \rho_0 \, \underline{v}_L = \rho_0 V \underline{U}_L \; , \tag{3.9}$$

$$\underline{u}_{sL} := \frac{1}{V} \int_V d^3x \; \underline{v}_L(\underline{a}(\underline{x},t),t) = \frac{1}{V} \int_V d^3a \; \frac{\rho_0}{\rho} \, \underline{v}_L(\underline{a},t) = \frac{1}{V} \int_V d^3a \; [1$$
$$- \frac{\rho - \rho_0}{\rho_0} + \dots] \; \underline{v}_L = \underline{U}_L - \frac{1}{\rho_0 V} \sum_{\underline{k}} n_{\underline{k}} \, \hbar\underline{k} + \dots \; . \tag{3.10}$$

Note that in (3.10) the spatial average is again taken in <u>local</u> coordinates $\underline{x}$, and use has been made of (2.8). Eq. (3.9) is exact. Together with (3.10) it shows that <u>a Lagrangian phonon has no momentum, but gives a contribution</u> $- (1/\rho_0 V)\hbar\underline{k}$ <u>to the average velocity.</u>

4. TWO-FLUID EQUATIONS

We consider liquid He II at temperatures below $0.6\,^{\circ}$K, where the thermal excitations consist entirely of phonons. Since their wavelengths are large compared to the interatomic distance they can be obtained in a good approximation from a continuum theory, i.e. from hydrodynamics, and our results of Section 3 can be applied.

A phonon has an energy $\varepsilon_{\underline{k}} = \hbar c \, |\underline{k}|$ if the background medium is at rest. In a frame in which the background is moving with a velocity $\underline{u}_s$, the wave vector is the same, $\underline{k}' = \underline{k}$, whereas the phonon energy is $\varepsilon_{\underline{k}}' = \varepsilon_{\underline{k}} + \hbar\underline{k} \, \underline{u}_s$. This follows from the Doppler shift formula, $\omega_{\underline{k}}' = \omega_{\underline{k}} + \underline{k} \, \underline{u}_s$, and is independent of whether a momentum is associated with a phonon or not.

The <u>crucial point</u> now is that $\underline{u}_s$, the velocity of the background, is to be identified with the <u>translational velocity</u> $\underline{U}$

of the background medium plus the average velocity due to the presence of the other phonons (if they give a contribution), i.e. with $\underline{u}_{sE}$ (3.5) for the Eulerian and $\underline{u}_{sL}$ (3.10) for the Lagrangian case.

Since $\sum_{\underline{k}} n_{\underline{k}} \varepsilon_{\underline{k}}'$ and, in a liquid, $\sum_{\underline{k}} n_{\underline{k}} \underline{k}$ are constants of the motion the mean phonon number in thermal equilibrium is given by $\langle n_{\underline{k}} \rangle = [\exp[(\varepsilon_{\underline{k}} + \underline{u}_s \hbar\underline{k} - \underline{u}_n \hbar\underline{k})/k_B T] - 1]^{-1}$, where $\underline{u}_n$ is interpreted as the drift velocity of the phonon gas. The mean wave vector density is

$$(1/V) \sum_{\underline{k}} \underline{k} \langle n_{\underline{k}} \rangle = (2\pi)^{-3} \int d^3k \; \underline{k} \langle n_{\underline{k}} \rangle =: (1/\hbar) \rho_n (\underline{u}_n - \underline{u}_s) , \qquad (4.1)$$

which defines ρ_n , the so-called density of the normal component in the two-fluid model.

The two-fluid equations are just the balance equations (always written in local coordinates) for the following conserved quantities: (i) energy of the phonons, (ii) wave vector of the phonons, (iii) momentum of the liquid as a whole, (iv) mass.

To deal with (i) and (ii) one can use the Boltzmann equation [3],[4] for the phonon distribution function $f(\underline{x},\underline{k},t)$

$$\partial f/\partial t + \dot{\underline{x}} \cdot \partial f/\partial \underline{x} + \dot{\underline{k}} \cdot \partial f/\partial \underline{k} = (\partial f/\partial t)_{coll} , \qquad (4.2)$$

where $\dot{\underline{x}} = \partial \omega_k/\partial \underline{k} + \underline{u}_s$ is the group velocity and $\dot{\underline{k}} = - (\partial \omega_k/\partial \rho) \underline{\nabla} \rho - \underline{\nabla} (\underline{u}_s \underline{k})$ can be found on purely kinematic grounds from the geometric meaning of the wave vector [2],[8]. In order to obtain the balance equations for (i) and (ii) one multiplies (4.2) by $\varepsilon_{\underline{k}}'$ and $\underline{k}$, respectively, and integrates over $\underline{k}$ [3],[4]. No $\underline{k}$ difference arises between Eulerian and Lagrangian phonons, and the resulting two-fluid equations are the same.

In order to deal with (iii) the expression for the momentum density is needed. In Euler's case it is according to (3.4),(3.5),(4.1)

$$\rho_0 \underline{U}_E + \rho_n (\underline{u}_n - \underline{u}_s) = \rho_0 \underline{u}_s + \rho_n (\underline{u}_n - \underline{u}_s) = \rho_n \underline{u}_n + \rho_s \underline{u}_s , \qquad (4.3)$$

where $\rho_s := \rho_0 - \rho_n$ is the so-called density of the superfluid component. For Lagrange's case the momentum density is according to (3.9),(3.10),(4.1) given by $\rho_0 \underline{U}_L = \rho_0 \underline{u}_s + \rho_n (\underline{u}_n - \underline{u}_s) = \rho_n \underline{u}_n + \rho_s \underline{u}_s$, i.e. again expression (4.3). This leads in both cases to the equation [4]

$$\partial (\rho_n \underline{u}_n + \rho_s \underline{u}_s)/\partial t + \underline{\mathrm{Div}} \; \boldsymbol{\Pi} = 0 , \qquad (4.4)$$

where $\boldsymbol{\Pi}$ is the momentum flux tensor.

Finally, the conservation of mass (iv) is described by the
continuity equation. For a system of massive particles, mass
current density and momentum density are the same thing, so
that in either case the continuity equation reads

$$\partial \rho / \partial t = - \underline{\nabla} \cdot (\rho_n \underline{u}_n + \rho_s \underline{u}_s) . \tag{4.5}$$

Thus the whole set of two fluid equations is shown to be identical for Eulerian and Lagrangian phonons. For the reversible and
linearized equations this has been shown by a slightly different method in [8].

References

1. H. Lamb: Hydrodynamics, Cambridge Univ. Press, London/
   New York, 1957
2. R. Kronig, Physica 19, 535 (1953)
3. J. de Boer: In Liquid Helium, ed. by G. Careri, Academic
   Press, New York, 1963, p. 1
4. I.M. Khalatnikov: Introduction to the Theory of Superfluidity, Benjamin, New York, 1965
5. W. van Saarloos: Physica 108A, 557 (1981)
6. A. Thellung: In Physics of Phonons, ed. by T.Paszkiewicz,
   Springer Lecture Notes in Physics, Vol.285 (1987) p.208
7. R. Kronig and A. Thellung: Physica 18, 749 (1952)
8. A. Thellung: Annals of Physics 127, 289 (1980)

Part IV

Quantum Evaporation

Long Lifetime Excitations in ^{4}He

A.F.G. Wyatt

Department of Physics, University of Exeter, Stocker Road, Exeter EX4 4QL, UK

1. INTRODUCTION

The discovery of quantum evaporation has opened up the possibility of studying excitations in liquid He which have long lifetimes. This arises because quantum evaporation [1,2] enables the excitations to be detected. High energy excitations are difficult or impossible to detect in the liquid but if they are converted into atoms in a known way then measurements can be made on them. Furthermore the evaporation process is only sensitive to excitations with energy $>$ 7.16 K and so it automatically selects out the interesting high energy excitations. It is particularly useful to be able to avoid low energy phonons.

The excitations in liquid ^{4}He have been seen by neutron scattering many times, for example [3,4]. The single excitations are phonons and rotons. The rotons can be further divided into two groups, those with positive or negative group velocities which we call respectively R^+ and R^- rotons. As we are interested in long lived excitations we need to observe them over long times. The excitations must also be propagating to eventually reach the free surface of the liquid and so they must have finite group velocities. This means we shall be concerned with rotons away from the roton minimum which is in contrast to thermodynamic measurements which are always dominated by rotons at the minimum energy, together with low energy phonons.

At high temperatures, T $>$ 1 K, the rotons and phonons are scattered by thermal rotons. The lifetimes are short and can be measured as a linewidth of the inelastic neutron scattering peak. The very best that a triple axis spectrometer can do is a line width of 0.1 K [5]. The neutron spin echo technique is more sensitive and can detect linewidths down to 10^{-2} K [5]. Using this technique it has been shown that the roton linewidth varies as $T^{1/2}e^{-\Delta/T}$, ie proportional to the thermal roton density [6]. Also the behaviour of phonons with high q is similar [5].

As the temperature of the liquid He is decreased the ratio of phonons to rotons in thermal equilibrium decreases from equality at 1.2 K to only 5 % rotons at 0.6 K. Below 0.5 K the rotons can essentially be ignored and injected phonons and rotons are scattered by thermal phonons. As the number of thermal phonons decreases as T^3, the scattering rapidly diminishes and it is at temperatures around 0.1 K and below where we expect long lived excitations.

If we can measure an excitation travelling a distance ℓ without being scattered then we can put a lower limit on its lifetime, $\tau > \ell/v$, where v is its group velocity. For typical values: $\ell \sim$ 1 cm, v $\sim 10^2$ ms^{-1} then $\tau = 10^{-4}$s. This is $> 10^5$ longer than can be detected with the best neutron techniques.

Springer Series in Solid-State Sciences, Vol. 79 **Elementary Excitations in Quantum Fluids**
Editors: K. Ohbayashi · M. Watabe © Springer-Verlag Berlin, Heidelberg 1989

2. MEASUREMENT TECHNIQUES

As hinted at above, time of flight techniques must be used to observe long
lived excitations. The situation is unusual as there are two parts to the
time between the injection of excitations and the detection of atoms.
There is the time of flight of the phonon or roton through the liquid and
then the time of flight for the atom through the vacuum. The atom part of
the time is essentially calculated as there is no question about free atom
behaviour. Of course it is important to work at sufficiently low
temperatures so that the atoms are not scattered by thermally excited atoms.

Scattering of the excitations in the liquid can show up in different
ways depending on the experimental arrangement, the type of scattering and
where it occurs. It can spread out the signal in time. This will occur if
the path length is increased and if the excitation energy is changed so
that its group velocity and the atom's velocity is changed. Scattering can
also attenuate the received signal. This will occur if the energy of the
excitation is reduced so that it cannot evaporate an atom and also if the
excitations are initially well collimated and scattering takes them out of
the beam.

The experimental arrangement can be simple with a thin film heater in
the liquid and a superconducting bolometer in the vacuum. The heater
injects a broad spectrum of phonons and positive group velocity rotons into
the liquid. These are collimated into a beam directed at the liquid's
surface. There, those excitations with $\hbar\omega > E_B$ (7.16 K) have a probability
of evaporating atoms. In practice, phonons must have a higher energy $\omega > \omega_c$
($\sim$ 9.5 K) to reach the surface, otherwise they are strongly scattered by
the 3 phonon process. The evaporated atoms can be collimated too. The
bolometer is naturally covered with a layer of liquid He which provides a
good receiving surface for the incident atoms. The sticking coefficient is
essentially unity. The bolometer responds to the condensation energy and
kinetic energy of the atom.

The direction of propagation can be usefully reversed. An atom beam can
be produced in the vacuum directed at the liquid surface. The atoms
condense and produce excitations [7]. However rotons produced in this way
must be reflected back to the surface to be detected by quantum evaporation
as they cannot be detected by a fast bolometer in the liquid He. We have
recently shown the existence of long-lived negative group velocity rotons
in this way.

3. EVIDENCE FOR LONG LIVED EXCITATIONS

In this section we review the evidence for long lived phonons and rotons.
There are three indications that excitations are not scattered in a
propagation experiment. The first is the fastest time between injection of
excitations and detection of the atoms. Any scattering will increase this
time. The second is the angular dispersion of atoms which can only exist if
the excitations can be collimated into a beam. The third is to account for
the received pulse shape assuming ballistic propagation. The pulse shape
however also depends on the spectrum of excitations created and this is not
always known. In practice the three indications are not separate but they
can be identified in the examples that follow.

At normal incidence the phonon-atom signal can be measured as a function
of the liquid depth, with the overall distance between heater and bolometer

constant. The time to the start of the signal varies due to the different dispersion for phonons and atoms. The heater creates a spectrum of phonons and for low liquid depths high energy phonons give the fastest signal as although the phonon group velocity is low, the atom has a longer path length and its velocity is higher. As the liquid depth increases, the phonon velocity becomes more important and so lower energy phonons give the fastest signal. If we use the leading part of the phonon–atom signal we can select the phonon energy by choosing the liquid depth. The good agreement between the measured and calculated times [1] confirms the above picture and that the phonons are ballistic.

The most direct qualitative evidence for ballistic propagation comes from the angular dispersion of the atom beam. Besides energy being conserved in quantum evaporation so is the component of momentum parallel to the liquid surface.

As a roton has much higher momentum than a phonon, for a given angle of incidence the atom from the roton comes off at a larger angle to the normal than the atom from the phonons. For an angle of incidence of $15°$ this allows a complete separation of the phonon–atoms and the R^+– atoms [2]. If either the roton or the phonon is scattered, the collimation would break down and the angular separation would not occur. This in fact happens with high power heater pulses [8]. The high number of excitations created causes mutual scattering. The excitation–atom signals are then slow and diffuse and have lost the angular dispersion seen at low powers.

The roton–atom pulse shape at low powers contains more information than the phonon–atom signal. This is because the roton–atom pulse is more dispersed in time due to the highest occupation of the roton states being at the lowest group velocities. This is contrary to the situation for phonons. The roton–atom signal pulse shows a significant tail due to the relatively large number of slow moving rotons. The pulse shape can be modelled assuming both a thermal distribution of rotons and ballistic propagation [8]. The time to the start of the pulse also agrees with calculation.

The dispersed roton–atom signal from ballistic rotons opens the possibility of roton spectroscopy. If the input pulse is short, then each point on the time axis of the received pulse can be identified with a roton energy. So if rotons of different energy are scattered by different amounts, this will appear as a change in the shape of the received pulse.

We were unable to detect any signal due to negative group velocity rotons (R^-) evaporating atoms. A recent experiment has shown that this is due to a heater not creating R^- rotons. However as mentioned above they can be created by condensing atoms and we have shown that they can travel 14 mm through the liquid He [9].

The experiments described above lead us to the conclusion that high energy phonons and rotons can propagate ballistically over distances of the order of 1 cm and therefore their lifetimes are at least 10^{-4} S. In the next section we discuss how scattering can now be studied.

4. SCATTERING BY THERMAL PHONONS

Now that beams of high energy excitations can be created and detected it should be possible to do quite detailed scattering experiments. However

the most obvious and simplest to do is the scattering by thermal phonons
and this we have done first. The experimental arrangement has a tightly
collimated beam at normal incidence. This allows the phonon–atom signal to
be easily distinguished from the broad roton–atom signal [10]. As the
temperature of the liquid He is raised between T = 0.1 and 0.2 K the
phonon–atom signal rapidly drops. The roton–atom signal only decreases a
little in the same temperature range.

The high energy phonons have to travel ~ 4 mm from the heater to the
free liquid surface. If at any point along the path a phonon scattering
event leaves the phonon energy $\omega < \omega_C$, that phonon rapidly decays to low
energies and will not be detected. The phonons injected with $\omega > \omega_C$ will
interact with the thermal phonons in a 4 phonon process [11]. The injected
phonon energy is some 2 orders of magnitude greater than the thermal
phonons at 0.1 K and so the change in the energy of the high energy phonon
is only ~ 1 % at each scattering event. The change in angle is also small,
~ 1°.

The scattering mean free path (λ) can be estimated from Khalatnikov's
expression for the scattering time [11]. For a phonon with ω = 11 K at
T = 0.1 K, λ = 76 μm. λ varies as $T^{-3} q^{-4}$, where T is the liquid He
temperature and q is the wave vector of the high energy phonon. Such a
phonon will therefore undergo many scatterings over a 4 mm distance.

At each scattering the energy of a phonon can increase or decrease a
little and so the phonon energy does a 1 dimensional random walk along the
energy axis. There is a chance that this walk will take the energy below ω_C
where the phonon is then lost. The probability of this happening will
depend on both the number of scattering events and how far away the
injected phonon energy is from ω_C. It turns out that the higher the
injected phonon energy the more chance it has of evaporating an atom.

A calculation has been made of the probabilities of various phonon
energies reaching the liquid surface. If the injected phonon spectrum is
taken to be Planck like with a charcteristic temperature of 0.7 K [12] the
spectrum of phonons incident on the surface can be calculated. Finally the
atoms signal can be found [13].

The calculated phonon–atom signal shows a rapid fall off with
temperature which is similar to the measured behaviour. However the
calculated change occurs at a somewhat lower temperature. This could be
due to the assumed 4 phonon scattering being too strong. The effect of the
change in angle has not yet been analysed but it too will decrease the
signal as the temperature is raised. Although all the questions have not
been settled, it does look promising that we can study weak scattering
processes with high energy phonons and rotons.

5. CONCLUSIONS

The combination of quantum evaporation and time of flight measurements has
enabled long–lived high–energy phonons and rotons to be detected. They can
propagate ballistically over distances ~ 1 cm in liquid He at low enough
temperatures. This now allows the scattering of these excitations to be
studied. The first study is the scattering of high energy phonons by
thermal phonons whose energy is ~ 2 orders of magnitude lower. This should
lead to a quantitative understanding of the 4 phonon process which hitherto
has been inaccessible.

REFERENCES

1. M.J. Baird, F.R. Hope and A.F.G. Wyatt: Nature **304**, 325, (1983)
2. F.R. Hope, M.J. Baird and A.F.G. Wyatt: Phys. Rev. Lett. **52**, 1528 (1984)
3. A.D.B. Woods and R.A. Cowley: Reports in Progress in Physics **36**, 1135 (1973)
4. W. Stirling: 75th Jubilee Conference on Helium-4, ed. by J.G.M. Armitage (World Scientific Pub. Co., Singapore 1983) p.109
5. F. Mezei and W.G. Stirling: 75th Jubilee Conference on Helium-4, ed. by J.G.M. Armitage (World Scientific Pub. Co., Singapore 1983) p.111
6. F. Mezei: Phys. Rev. Lett. **44**, 1601 (1980)
7. D.O. Edwards, G.G. Ihas and C.P. Tam: Phys. Rev B **16**, 3122 (1977)
8. M. Brown and A.F.G. Wyatt: Proceedings of LT18, Kyoto, 1987 p.385
9. G.M. Wyborn and A.F.G. Wyatt: Proceedings of LT18, Kyoto, 1987 p.2095

10. M.J. Baird, B. Richards and A.F.G. Wyatt: Proceedings of LT18, Kyoto, 1987 p.387
11. I.M. Khalatnikov: An Introduction to the Theory of Superfluidity, (W.A. Benjamin Inc, New York, Amsterdam, 1965)
12. A.F.G. Wyatt, R.A. Sherlock and D.R. Allum: J. Phys. C Solid State Physics **15**, 1897 (1982)
13. A.F.G. Wyatt: Proceedings of LT18, Kyoto, 1987 p.7

Part V

Vortices, Ripplons and Ions

Vortex Excitations and the Superfluid λ Transition

G. A. Williams

Department of Physics, University of California, Los Angeles, CA 90024, USA

The role of vortex excitations in the critical region near the superfluid λ transition of ^{4}He is discussed. A simple model using circular vortex rings has been constructed following the general ideas of Kosterlitz and Thouless, extended to the case of three dimensions. A real-space renormalization technique generates a power-law phase transition that satisfies the Josephson scaling relation and has the correct correlation length. The physical picture that emerges agrees with the original intuitive proposals of Feynman and Onsager, and with recent Monte Carlo simulations.

1. INTRODUCTION

It has been known for a number of years that vortex rings are elementary excitations of superfluid helium, and can be thermally excited on just the same basis as the more familiar phonons and rotons. A well-known physical consequence of such thermal vortex excitations is the slow decay of persistent currents [1]. In their original papers proposing quantized circulation, both ONSAGER [2] and FEYNMAN [3] proposed that vortex excitations might be responsible for the superfluid λ-transition, but this has proven to be a much more controversial proposal. A number of authors have reiterated this idea (see refs. 4 and 5 for a partial list), but actual calculations did not get very far in predicting the properties of the phase transition. In the meanwhile, perturbation techniques such as the high-temperature expansion and the 4-ϵ expansion [6] were developed which were able to successfully calculate the critical exponents of the transition. These expansions are very formal and mathematical, however, and give little physical insight into the nature of the transition. The momentum-space representation that is used makes the identification of elementary excitations very difficult. A localized object such as a vortex ring becomes spread over a large region of k space when Fourier transformed.

The role of vortices in phase transitions was first demonstrated by BEREZINSKII [7] and KOSTERLITZ and THOULESS [8] for the case of two dimensions (2D). They showed that the transition in that case could be understood in great detail by considering the excitation of vortex pairs of equal and opposite circulation. Experiments in thin helium films have well verified this theory [9]. It is interesting to note that the 2D transition was first explored theoretically [10] using a high-temperature expansion that made no mention of vortices. The expansion was difficult to use, and little progress on the physics of the transition was made until Kosterlitz and Thouless introduced their vortex pair model.

The role of vortices in 3D transitions has not been as well confirmed as in the 2D case, and some physicists remain skeptical that they contribute at all. However, recent Monte Carlo simulations [11] have given strong

Springer Series in Solid-State Sciences, Vol. 79   **Elementary Excitations in Quantum Fluids**
Editors: K. Ohbayashi · M. Watabe   © Springer-Verlag Berlin, Heidelberg 1989

evidence that vortices are in fact responsible for the transition. The
simulations were able to vary the core energy of the vortex excitations.
Increasing this energy suppresses the thermal activation of the vortices,
and the T_c of the transition was found to shift rapidly towards infinite
temperature as the vortex density was decreased.

2. VORTEX RING MODEL

We have recently introduced a simple model of the λ transition using
circular vortex rings as the starting point for a real-space renormalization
calculation [5]. The model is still speculative and does not give accurate
values for the critical exponents, but it is still felt that the underlying
ideas provide a physical basis for understanding the transition.

The energy of a large circular vortex ring of radius R is given by

$$U_o = 2\pi^2 \frac{\hbar^2}{m_4^2} \rho_s^o R \left(\ell n \frac{R}{a} + c \right) . \tag{1}$$

Here m_4 is the He atom mass, ρ_s^o is the unrenormalized superfluid density,
$c = 0.464$ is a constant related the core energy, and a is the vortex core
radius, related to the ϕ^4 coupling constant of the Landau-Ginzburg-Wilson
Hamiltonian [12]. It is useful to make use of a magnetostatic analogy to
treat these rings as dipole current loops. In an applied flow field these
dipole loops tend to orient perpendicular to the flow, and the backflow from
the oriented rings will reduce the applied flow. Since the superfluid
density ρ_s is proportional to the net resulting current, this corresponds to
a reduction of ρ_s. This screened density can be calculated using linear
response theory, defining a permeability $\mu(R) = \rho_s^o/\rho_s(R) = 1 + 4\pi\chi(R)$ where
now $\rho_s(R)$ is the scale-dependent density seen by a ring of radius R. The
susceptibility χ can be calculated by multiplying the polarizibility
of a ring ($\sim \rho_s^o R^4$) and the number density of thermally excited rings
($\sim R^2 e^{-U_o/k_B T}$) and integrating over R, yielding an equation for ρ_s:

$$\frac{1}{\rho_s(R)} = \frac{1}{\rho_s^o} + A \int_{a_o}^{R} \left(\frac{R}{a_o} \right)^6 e^{-U_o/k_B T} \frac{dR}{a_o} . \tag{2}$$

Here a_o is the "bare" core size and A is a constant whose exact value is
unimportant. This equation is the starting point for a real-space
renormalization calculation [5] using the technique of JOSÉ et al [13]. The
core size a_o is increased by a small amount $a_o \to a_o + \delta$, and the physics is
required to remain unchanged by the change in scale. The result of this
technique is that Eq. (2) is recovered, but with the change that the vortex
energy in the exponential is replaced by a screened value $U_o \to U(R)$, where

$$U(R) = \int_{a_o}^{R} \frac{1}{\mu(R)} \frac{\delta U_o}{\delta R} dR . \tag{3}$$

There is also a <u>replacement</u> of the core radius a in Eq.1 with an effective
radius $a_r = a_o \sqrt{\rho_s^o/\rho_s(R)}$, which grows as the transition is approached. The
expression for the screened energy is the same form as originally proposed
by Kosterlitz and Thouless, and results from the interaction between a large
ring of radius R and the rings of smaller size in its vicinity. Similarly,
the effective core size can probably be viewed as arising from an excess
density of smaller rings near the core of the larger ring.

The screened version of Eq.2 can be solved by iteration, starting with the bare value ρ_s^o at R = a_o, and then recursively solving Eqs. 2 and 3 out to R → ∞. The result is that ρ_s is found to drop to zero at a critical temperature T_c with a power-law behavior, $\rho_s \sim (T-T_c)^\upsilon$, where υ is found from numerical iterations to be 0.53. T_c is determined by a critical value of the product $\rho_s^o\, a_o$, and hence depends on the microscopic parameters of the system. The model can be shown to obey the Josephson scaling relation, that the superfluid exponent υ and the specific heat exponent α are related by $\upsilon = (2-\alpha)/3$. The coherence length of the model,

$$\xi = \frac{m_4^2\, k_B T}{\hbar^2\, \rho_s} , \qquad (4)$$

coincides with the known result [6]. This length can now be identified as the diameter of the largest ring appreciably excited at a given value of $T-T_c$. At T_c where $\rho_s = 0$, rings of infinite size are generated, made possible by the screening effects.

3. <u>DISCUSSION</u>

The above model is offered as a speculative description of the underlying physical mechanism of the λ transition. The fact that the superfluid exponent does not match the known value $\upsilon = 0.67$ certainly indicates that the model has defects, or is incomplete. The use of strictly circular rings is a major simplifying assumption. The actual vortex loops excited will be deformed and wiggly, with considerable entropy in the deformations [14]. Some progress in calculating the properties of wiggling "strings" has been made [15] in a related problem, that of the phase transition of He films in porous materials [4, 16]. It may be possible to incorporate similar line fluctuations in the ring model.

A scenario of the λ transition can be postulated, based on the ring calculation and the Monte Carlo simulations. At low temperatures ρ_s^o is large, and hence U_o from Eq.1 is large and unfavorable for the thermal excitation of rings. The phonons and rotons are the dominant excitations, and determine the temperature dependence of ρ_s^o as in the Landau two-fluid model. At high temperatures near the λ point, however, ρ_s^o drops to small values, and it then becomes favorable to excite rings of radius larger than a_o. These rings lower the superfluid density to the renormalized value ρ_s, which allows rings of even larger size to be excited. With increasing temperature this process runs away, resulting in $\rho_s = 0$ at T_λ. Vortices are the "soft" excitation mode in helium, since the phonon and roton energies are known to be finite above T_λ.

A puzzle remains about how the Bose condensed zero-momentum fraction n_o fits into this description of the λ transition, in which quantized vortices exist above as well as below T_λ. It is thought that the condensate fraction falls to zero at the transition [17], but how the vortices might bring this about is unclear. The vortices certainly disrupt the long range order for T > T_λ, and this may somehow preclude the formation of a macroscopic condensed state. The relationship between n_o and ρ_s has in fact never been established in any of the models of the λ transition. The perturbation expansions [6] and the present vortex model require only the existence of a two-component order parameter to describe the critical properties of the transition, and make no reference at all to the condensate fraction. Further work will be needed to make this connection to the more microscopic description of helium.

ACKNOWLEDGEMENT

This work was supported by the U.S. National Science Foundation, grant DMR 84-15705.

REFERENCES

1. J.S. Langer and J.D. Reppy, in *Progress in Low Temperature Physics*, ed. by C.J. Gorter (North-Holland, Amsterdam, 1970) Vol.6, p.1
2. L. Onsager, Nuovo Cimento Suppl. **6**, 249 (1949)
3. R.P. Feynman, in *Progress in Low Temperature Physics*, ed. by C.J. Gorter (North-Holland, Amsterdam, 1955) Vol.1, p.17
4. V. Kotsubo and G.A. Williams, Phys. Rev. B**33**, 6106 (1986)
5. G.A. Williams, Phys. Rev. Lett. **59**, 1926 (1987)
6. P. Hohenberg, Physica **109** & **110B**, 1436 (1982)
7. V. Berezinskii, Zh. Eskp. Teor. Fiz. **59**, 907 (1970) [Sov. Phys. JETP **32**, 493 (1971)]
8. J.M. Kosterlitz and D.J. Thouless, J. Phys. **C6**, 1181 (1973)
9. I. Rudnick, Phys. Rev. Lett. **40**, 1454 (1978); D.J. Bishop and J.D. Reppy, Phys. Rev. **B22**, 5171 (1980)
10. H.E. Stanley, Phys. Rev. Lett. **20**, 150 (1968)
11. G. Kohring, R. Schrock, and P. Wills, Phys. Rev. Lett. **57**, 1358 (1986)
12. P.H. Roberts and J. Grant, J. Phys. **A4**, 55 (1971)
13. J.V. José, L.P. Kadanoff, S. Kirkpatrick and D.R. Nelson, Phys. Rev. **B16**, 1217 (1977)
14. C.F. Barenghi, R.J. Donnelly, and W.F. Vinen, Phys. Fluids **28**, 498 (1985)
15. F. Gallet, private communication.
16. T. Minoguchi and Y. Nagaoka, Jpn. J. of App. Phys. Suppl. **26-3**, 327 (1987); R. Guyer and J. Machta, private communication.
17. E.C. Svensson, in *75th Jubilee Conference on Helium 4*, ed. by J.G. Armitage (World Scientific, Singapore, 1983) p.10

Inhibition of Vortex Nucleation by Phonons in He II

P.C. Hendry[1], N.S. Lawson[1], C.D.H. Williams[1], P.V.E. McClintock[1],
and R.M. Bowley[2]

[1]Department of Physics, University of Lancaster, Lancaster, LA1 4YB, UK
[2]Department of Physics, University of Nottingham, Nottingham, NG7 2RD, UK

1. Introduction

The superfluidity of ^{4}He can in principle break down in two quite distinct ways when
an object moves through the liquid, or the liquid flows through a tube, at a sufficiently
high velocity. Either rotons may be produced above the Landau critical velocity [1,2]
or, alternatively, quantized vortices may be created [3,4] or expanded [5]. Because He II
invariably contains at least a few pinned vortices [6], the phenomenon of vortex *creation*
is extremely difficult, though perhaps not entirely impossible [7], to observe in flow
experiments. The creation process can, however, be investigated very readily through
studies of the motion of negative ions which, in view of their small size (of radius $\sim$
1 nm), are most unlikely to be affected by remanent vorticity [6]. In this chapter we
report and discuss an unexpected temperature dependence of the rate at which vortices
are nucleated by negative ions for pressures $P > \sim 15$ bar.

2. Experimental Details

The vortex nucleation rate ν was measured by means of the electric induction technique
described previously [4], but with the ion source modified [2] so as to minimise vortex
creation close to the field emitters which, otherwise, would severely have curtailed the
magnitude of the bare ion current entering the induction space at low pressures. As in
the earlier work [4], the speed of the ions was controlled by a balance between the force
due to the applied electric field and the momentum loss associated with roton emission.
The ^{4}He sample was isotopically purified [8] in order to eliminate the influence of ^{3}He
isotopic impurities which, for natural helium below 0.5K, can be very pronounced and
leads to enormously enhanced nucleation rates [9]. The experimental cell was the one
used previously for Landau velocity measurements [2], but with the electrode geometry
modified [4] to enable ν to be measured. It was mounted in a simple dilution refrigerator
that enabled the temperature to be stabilized within $50 < T < 500$mK.

3. Experimental Results

Some typical experimental measurements of the temperature dependence of ν are plotted for three pressures in Fig. 1. Because the absolute value of ν is very strongly
dependent on pressure [4], the electric field has also been varied between plots so as to
bring them closer together and thus display to better advantage the striking qualitative
change that occurs in the form of $\nu(T^{-1})$ as P is increased. In particular, it is found
that a distinct local minimum in $\nu(T^{-1})$ develops as the pressure is increased above $\sim$
15 bar. In Fig. 2, $\nu(T^{-1})$ is plotted (data points) for three electric fields at the same
pressure with a linear ordinate scale; it is evident that the position of the minimum
is weakly dependent on electric field. The drift velocities $\bar{v}$ of the ions were measured
from the same signals as those from which ν was extracted and it was found that, in the
region of the local minimum, $\bar{v}(T)$ remained constant within the experimental precision
of $\pm \sim 0.2\%$: we return to this point below.

Springer Series in Solid-State Sciences, Vol. 79   **Elementary Excitations in Quantum Fluids**
Editors: K. Ohbayashi · M. Watabe   © Springer-Verlag Berlin, Heidelberg 1989

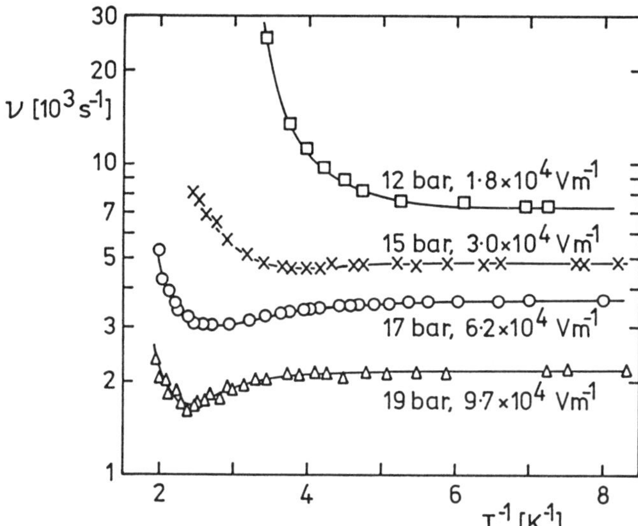

Figure 1. The vortex nucleation rate, ν, measured (points) as a function of reciprocal temperature, T^{-1}, at four pressures. The full curve at 12 bar is a fit of (1) to the data; those at the other pressures are guides to the eye

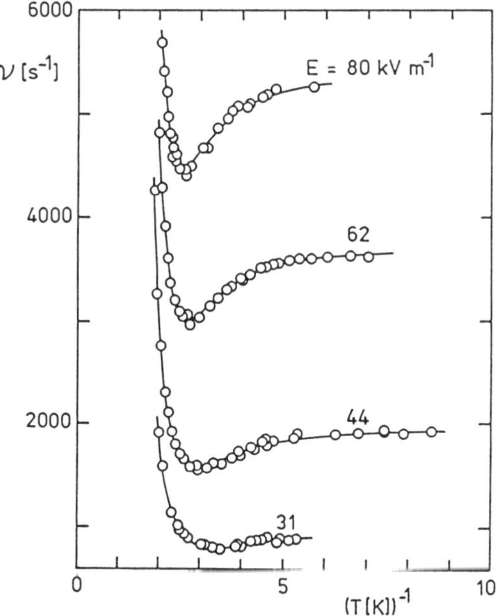

Figure 2. The vortex nucleation rate, ν, measured (points) as a function of reciprocal temperature, T^{-1}, at four electric fields under a pressure of 17 bar. The full curves are fits of (3) to the data

Although the present apparatus (designed for measurements of large ν values in low E at low T and under low P) cannot currently be used to establish whether or not local minima in $\nu(T^{-1})$ exist for $P > 19$ bar, we suspect that this is probably the case; the minimum temperature (0.3K) of the earlier work [4] probably lay very close to the positions of the minima, and there would have been no reason to anticipate the existence of a rise in ν with a further decrease of T.

4. Discussion

The $\nu(T^{-1})$ measurements at 12 bar (squares in Fig. 1) have been fitted (full curve) by a relation of the form

$$\nu = \nu_0 + Ae^{-\epsilon/k_B T} \tag{1}$$

with $\nu_0 = 7.5 \times 10^3 \text{s}^{-1}$, $A = 6.3 \times 10^8 \text{s}^{-1}$, $\epsilon/k_B = (3.1 \pm 0.1)$K where the latter quantity represents the weighted mean obtained by fitting (1) to several sets of data recorded at different electric fields. The limiting low temperature nucleation rate ν_0 may be interpreted [10] as corresponding to the (temperature-independent) rate at which macroscopic quantum tunnelling takes place through a small potential barrier impeding the nucleation process, as proposed by MUIRHEAD, VINEN and DONNELLY [11]. As the temperature rises above ~ 0.2K, the Arrhenius term in (1), corresponding to thermal activation over the barrier, becomes important and ν then rises rapidly with further increase of T. The experimental ϵ/k_B is in excellent agreement with the value estimated by Muirhead, Vinen and Donnelly, so that the measurements in question can be regarded as lending strong support to their model.

The local minima in $\nu(T^{-1})$ seen for pressures above ~ 15 bar were quite unexpected and their origin has yet to be accounted for. The simplest inference is that, for $P > \sim$ 15 bar, ν_0 decreases slightly with rising T before the effect of the Arrhenius term becomes dominant. The principal change occurring in the liquid, as the temperature rises through the range in question is, of course, the increase in the density of phonons; it seems reasonable, therefore, to guess that it is the phonons that in some way give rise to the minima in $\nu(T^{-1})$.

The most straightforward way in which phonons might affect ν is through their bringing about a slight reduction in the drift velocity $\bar{v}$ of the ions, corresponding to the momentum loss in ion-phonon scattering; an effect which would, of course, increase rapidly with increasing T. To allow for this possibility, it is convenient to define an effective electric field

$$E' = E - KT^3 . \tag{2}$$

In the temperature independent regime below 0.2K, $\nu_0 \propto E^2$ to a good approximation. If we suppose that $\nu_0 \propto E'^2$ in the range $T > 0.2$K, then it is straightforward to demonstrate that, to first order in KT^3/E, (1) should be replaced by

$$\nu = \nu_0 \left(1 - \frac{2KT^3}{E}\right) + Ae^{-\epsilon/k_B T} \tag{3}$$

to allow for the drag due to phonon scattering. This relation should provide a reasonable approximation provided $KT^3 << E$, but cannot be expected to apply outside this range. The full curves of Fig. 2 represent fits of (3) to the data with $K = 1.64 \times 10^5$ $\text{Vm}^{-1}\text{K}^{-3}$, and are clearly in good agreement thus apparently confirming the phonon scattering hypothesis.

The difficulty with this approach, however, as already mentioned in section 3, is that the measured values of $\bar{v}$ appear to be constant within experimental error in the temperature range under consideration. On the other hand, the absolute value of ν

186

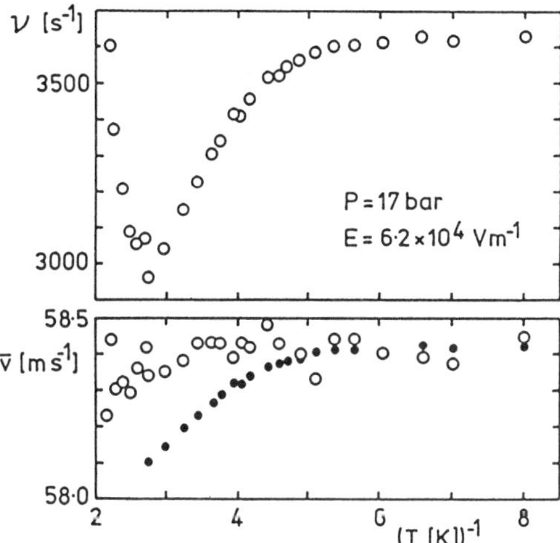

Figure 3. (a) The measured vortex nucleation rate, ν, plotted as a function of reciprocal temperature, T^{-1}, with an expanded ordinate scale, for the indicated values of P and E. (b) Ionic drift velocities $\bar{v}$ measured (open circles) for the same set of signals as were used for (a), compared with the values (filled circles) that would be needed to account for the measured nucleation rates if the local minimum in $\nu(T^{-1})$ were due to a reduction in $\bar{v}$ caused by ion/phonon scattering

is highly sensitive to small changes in $\bar{v}$, so it is necessary to establish whether the reduction in $\bar{v}$ corresponding to the dip in $\nu(T^{-1})$ would in fact be detectable in our experiments. A convenient consistency test is to make a plot of $\nu(\bar{v})$ from measurement of ν and $\bar{v}$ at a range of E in the temperature independent ($T < 0.2$K) range: the $\nu(T^{-1})$ values of data in the dip for $T > 0.2$K can then be associated with the values of $\bar{v}$ necessary to produce them, and comparison can be made to the actual values of $\bar{v}$ measured in the experiment. Such a comparison is shown in Fig. 3. Figure 3(a) plots $\nu(T^{-1})$ for one value of E at 17 bar. Fig. 3(b) plots the values of $\bar{v}$ corresponding to the same signals (open circles), compared with the values of $\bar{v}$ that would be needed to account for the dip (filled circles). It is evident that, although the fall in $\bar{v}$ needed to account for the observed fall in ν is very small, it lies well outside the scatter in the data. One may conclude, therefore, that the dip in $\nu^{-1}(T)$ does not arise simply from a slight reduction in the average ionic velocity due to phonon scattering, despite the excellence of the fits of (3) to the data of Fig. 2.

Nonetheless it should be noted that the dips in $\nu(T^{-1})$ could conceivably still arise from a reduction in velocity due to phonon scattering if the ion-phonon scattering cross section were strongly velocity dependent. In such a case, the high velocity tail on the velocity distribution function $f(v)$ [12] could be strongly attenuated without there being any measurable effect on $\bar{v}$ itself: since it is the ions in the tail that have the largest probability of creating vortex rings, ν is naturally very highly sensitive to any change in the shape of $f(v)$ in this region. We do not regard this scenario as very probable but, at present, it is a possibility that can neither be confirmed nor definitely excluded.

Even if the phonons do not significantly affect ν through changes in $\bar{v}$, or in $f(v)$, it remains possible that they may exert a direct inhibiting influence on the macroscopic quantum tunnelling process itself. Modulation by phonons of the height of the barrier would, of course, be expected to increase the tunnelling rate. The experimental situation is quite complicated, however: first, the phonons impinge at all angles on the ion so that those striking it in the direction perpendicular to v could act as an effective source of dissipation without either raising or lowering the barrier; secondly, even in the absence of phonons, the barrier height continuously decreases with time as the ion accelerates, a feature which could be important in determining the tunnelling rate and the influence of phonons upon it, but which has not yet been taken properly into account.

5. Conclusion

For pressures $P >\sim 15$ bar, there exists a temperature range within which an *increase* of temperature causes a *decrease* in the rate at which vortex rings are nucleated by negative ions moving through isotopically pure superfluid ^{4}He. This unexpected phenomenon is not due to a decrease in the ionic drift velocity caused by phonon scattering. It appears more likely that it results from a direct influence of phonons on the process of macroscopic quantum tunnelling through the potential barrier impeding nucleation events, but further work will be required to identify in detail the mechanism through which phonons are able to inhibit the tunnelling process.

References

1. L. Meyer and F. Reif: Phys. Rev. 123, 727 (1961)

2. T. Ellis and P.V.E. McClintock: Phil. Trans. R. Soc. Lond. A 315, 259 (1985)

3. G. W. Rayfield and F. Reif: Phys. Rev. 136, A1194 (1964)

4. R. M. Bowley, P.V.E. McClintock, F. E. Moss, G. G. Nancolas and P.C.E. Stamp: Phil. Trans. R. Soc. Lond. A 307, 201 (1982)

5. J. T. Tough: In Progress in Low Temperature Physics, ed. D. F. Brewer, Vol. VIII, p133 (North-Holland, Amsterdam, 1982)

6. D. D. Awschalom and K. W. Schwarz: Phys. Rev. Lett. 52, 49 (1984)

7. O. Avenel and E. Varoquaux: Phys. Rev. Lett. 55, 2704 (1985). E. Varoquaux, M. W. Meisel and O. Avenel: Phys. Rev. Lett. 57, 2291 (1986)

8. P. C. Hendry and P. V.E. McClintock: Cryogenics 27, 131 (1987)

9. G. G. Nancolas, R. M. Bowley and P.V.E. McClintock: Phil. Trans. R. Soc. Lond. A 313, 537 (1985)

10. P. C. Hendry, N. S. Lawson, C. D. H. Williams, P. V. E. McClintock and R. M. Bowley: in Proc. 18th Int. Conf. on Low Temperature Physics, ed. by Y. Nagaoka, Jap. J. Appl.Phys. (Suppl. 26-3) 26, 73 (1987)

11. C. M. Muirhead, W. F. Vinen and R. J. Donnelly: Phil. Trans. R. Soc. Lond. A 311, 433 (1984)

12. R. M. Bowley and F. W. Sheard: Phys. Rev. B 16, 244 (1977)

Ions and Electrons Trapped at the Surface of Superfluid Helium: Probes of the Study of Ripplons?

W.F. Vinen

Department of Physics, University of Birmingham, Birmingham B15 2TT, UK

Among the elementary excitations associated with a quantum fluid, if it has a free surface, are the quantized surface waves or *ripplons*. It is probably fair to say that less is known about the detailed properties and behaviour of these excitations than about the other excitations in quantum fluids. In this paper I want to speculate about the possibility of having a new experimental tool that may help in a small way to fill this gap.

The following are examples of experimental studies, already carried out, that are potentially relevant to our knowledge about the ripplons on the surface of superfluid ^{4}He.

1. The temperature dependence of the surface tension (see, for example, the reviews by EDWARDS & SAAM [1] and by EDWARDS [2]). At least at low temperatures this is believed to arise from the thermal excitation of ripplons, and there is good experimental evidence to support this view. Measurement of the surface tension gives the free energy of the ripplon gas. The free energy turns out to be consistent with the ripplons having at small wavenumbers a dispersion relation that is the same as that of classical capillary waves on an ideal fluid, although deviations must occur as the wavenumber approaches the reciprocal of the interatomic spacing.

2. Optical detection. Capillary waves, generated by a variety of techniques, can be detected by the diffraction of light at the liquid surface. Thermally-excited ripplons can also be detected in this way, provided that they have the appropriate wavelength (see, for example, [3]). The total scattered intensity from thermally excited ripplons has been studied, but not the spectrum [4].

3. The contribution of ripplons to heat conduction along a thin film has been observed and studied [5]. The ripplons play a role analogous to bulk excitations in counterflow heat conduction through a channel. It appears that the ripplons in the film suffer a drag due to momentum exchange with the solid substrate. For thick layers of helium this drag should be absent, but the contribution to thermal conduction in this case has not been observed. Similarly, the corresponding propagation of surface second sound has not been observed (this is true only in pure ^{4}He; in the presence of small concentrations of ^{3}He surface sound has been observed, but the dominant excitations are then ^{3}He atoms adsorbed at the surface and not ripplons).

4. Scattering of atoms, electrons, and neutrons from the surface [1,2]. There have been many studies of this type, some of which undoubtedly involve the ripplons, but they seem not so far to have yielded any reliable new information about the ripplons.

There remain then, as far as I am aware, many aspects of ripplons and the ripplon gas about which we have little or no experimental evidence. For example: the ripplon dispersion relation at high wavenumbers (for recent theoretical work see [6]); ripplon-ripplon interactions [7, 8], and the rate at which the ripplon gas comes into internal equilibrium; the ripplon-phonon interaction [9]; interaction between the ripplons and the walls of a containing vessel (the 'ripplon Kapitza resistance'). We ought therefore to be on the lookout for any new experimental tools that might throw light on any of these topics.

I want to speculate that the study of ions trapped below the surface of superfluid helium might provide such a tool. I can only speculate at this stage because the necessary theory has not yet been worked out properly, and because the relevant experiments are still at a relatively early stage of development. But I thought it might be of interest to draw attention to some possibilities. Much of the relevant background will be discussed in detail at the forthcoming international conference on low temperature physics (LT18), both by my own group and by the group of G. A. Williams at UCLA, so I shall keep the background to a minimum at this Symposium.

In the presence of an externally applied electric holding field (E_O) ions can be trapped just below the surface of the helium; the combination of E_O and the field due to the electrical image of the ion in the free surface produces a net potential with a minimum at a distance, z_O, below the surface that depends on E_O. The mobility of ions trapped in this way can be measured in various ways; at low temperatures the most convenient way is from the linewidth of two-dimensional plasma resonances in a pool of such ions. At higher temperatures this mobility is determined by collisions with the bulk excitations in the liquid, but at the lowest temperatures, typically below 50-100 mK, it is determined by collisions with ripplons. It is this fact that provides us, I believe, with a new tool for the study of the ripplons. Electrons can also be trapped at the surface of superfluid helium, and I shall return to this point later.

Experimental studies of the surface ionic mobility have been carried out by Gary Williams' group and by my own, and they will be described at LT18. In my own group we have tried to work out a theory of the ripplon-limited mobility. I will not go into the details here but simply say that an important process is probably the scattering of ripplons which have frequencies that resonate with the frequency of vertical vibration of the ions in the holding potential well. This frequency varies with the holding field E_O, and is typically 100 MHz, which corresponds to a ripplon wavelength of about 100 nm. In practice the mobility has always been measured at a finite frequency (of order 100 kHz), and it is possible that another important process is the direct generation of ripplons at this frequency. I have to admit that the agreement with experiment is not yet very good; the calculated mobility tends to be rather too high, and the observed mobility has sometimes a real or apparent dependence on plasma density that is not convincingly accounted for by the theory. However, I am not yet convinced that all the experimental results are reliable, and the theory undoubtedly requires some development. Nevertheless, I feel that the essential features of the theory may turn out to be good, and that it is therefore worth starting to speculate whether one might at some stage in the future use it with the observed surface ionic mobility to learn something about the ripplons.

For example, the mobility might be used as a ripplon thermometer. The ion plasma could constitute a rather good thermometer because it has a heat capacity that is much less than that of the ripplons. One could then try adding heat to the ripplons and seeing how long it takes the ripplon gas to cool, either by interaction with the phonons in the helium, or by loss of energy directly to the walls of the containing vessel. Hence we might obtain information about the ripplon-phonon interaction and about the ripplon Kapitza resistance. Existing theory suggests that the relevant time constants might be quite long and easily measurable, especially at very low temperatures. Given that the ions probe ripplons of a particular frequency (tunable by changing E_O), one could perhaps investigate the rate at which energy can be transmitted within the ripplon gas. By localizing the ions at different places on the liquid surface, one might be able to study the flow of heat through the ripplon gas. And one could study how all these things are affected by any ^{3}He adsorbed on the surface. The extent to which an assembly of ions can emit ripplons, or interact coherently with ripplons, will depend on the ripplon mean free path, so that a study of the ripplon limited mobility might lead to information about ripplon-ripplon interactions. And so on.

As is well-known, it is also possible to trap electrons at the surface of superfluid helium, but just above the surface and not just below it (see, for example, the review [10]). Two-dimensional electron plasmas formed in this way have been the subject of much experimental and theoretical study. At low temperatures the electron mobility is limited by ripplons, as is the case for the ions. In principle, one might use this electron system as another probe of the ripplons. A possible disadvantage is connected with the fact that in this case the ripplon limited mobility is severely modified at low temperatures by crystallization of the plasma, in a way that is complicated and not understood in complete detail. It is likely that crystallization of the ion plasma must also occur, but experimental evidence so far suggests strongly that it has little effect on the mobility. In any case the two systems probably interact with ripplons of different wavenumbers, and they may therefore complement each other as ripplon probes.

I emphasise that these are speculations. But I think that they stand some chance of being well-based and therefore of leading to some interesting new information about the behaviour of one of the excitations in a quantum fluid about which we do not yet know a great deal.

<u>References</u>

1. D. O. Edwards, W. F. Saam: In <u>Progess in Low Temperature Physics</u>, ed. by D. F. Brewer, Vol. VIIA, ch. 4, p. 283 (North Holland Publishing Co. Amsterdam, 1978)
2. D. O. Edwards: Physica B<u>109</u> & <u>110</u>, 1531 (1982)
3. S. T. Boldarev, V. P. Peshkov: JETP Letters <u>17</u>, 297 (1973)
4. F. Wagner: J. Low Temperature Phys. <u>13</u>, 317 (1973)
5. I.B.Mantz, D.O.Edwards, V.U.Nayak: Phys. Rev. Letters 44, 663 (1980)
6. E. Krotscheck, S. Stringari, J. Treiner: Phys. Rev. B<u>35</u>, 4754 (1987)
7. W. F. Saam: Phys. Rev. A<u>8</u>, 1918 (1973)
8. H. Gould, V. K. Wong: Phys. Rev. B<u>18</u>, 2124 (1978)
9. W. F. Saam: Phys. Rev. B<u>12</u>, 163 (1975)
10. A. J. Dahm, W. F. Vinen: Phys. Today <u>40</u>, 43 (1978)

Excitation Emission from Ions Moving at Supercritical Velocities in He II

C.D.H. Williams, P.V.E. McClintock, and P.C. Hendry

Department of Physics, University of Lancaster, Lancaster LA1 4YB, UK

1. Introduction

The superfluidity of He II is disrupted when an object moves through the liquid at a sufficiently high speed. Since superfluid flow is necessarily dissipationless the superfluidity is destroyed if, for example, conditions facilitate creation of excitations within the He II. Such processes may involve: creation of vortices [1,2]; evolution of an existent vortex structure [3] into a dissipative tangle; or, if the Landau critical velocity V_L is exceeded, the emission of rotons [4,5]. It is the emission of rotons by a moving object that we discuss below: the object in question being a negative ion which, as we shall see, can exist in several different forms.

2. Experimental details

The most reliable methods of determining ion velocities and mobilities are based on time-of-flight measurements. Figure 1 is the schema of such an experiment to investigate negative ions. The ions are produced near the top of the cell, usually by field emission [5] but special sources [6] are needed to create the 'fast' and 'exotic' [7 - 10] species of

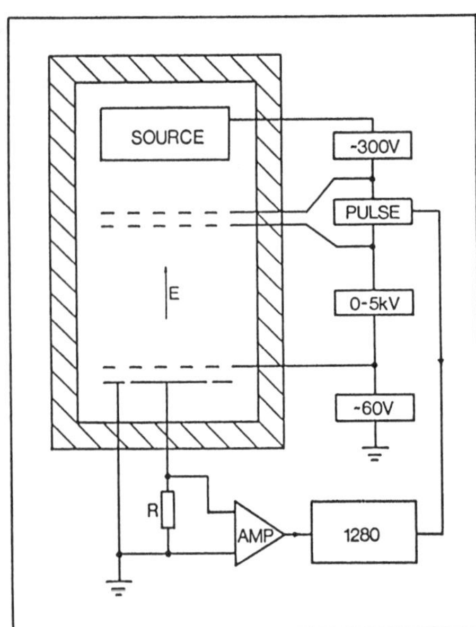

Figure 1. Schematic diagram of a typical experiment. The cell is constructed from Perspex. The Nicolet 1280 controls the experiment and performs signal averaging.

Springer Series in Solid-State Sciences, Vol. 79   **Elementary Excitations in Quantum Fluids**
Editors: K. Ohbayashi · M. Watabe   © Springer-Verlag Berlin, Heidelberg 1989

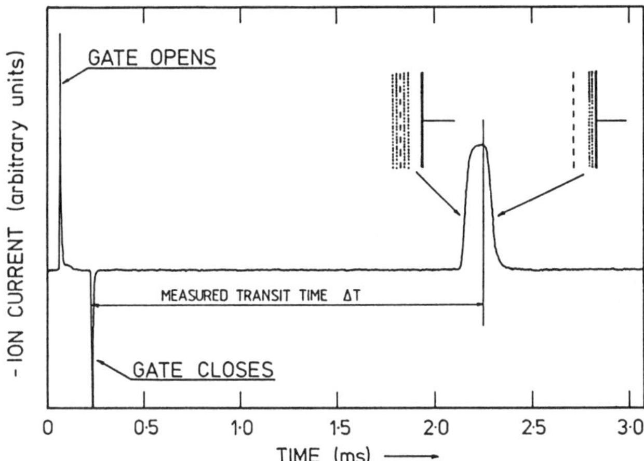

Figure 2. A typical ion current vs. time plot

ion. The ions are held in the gate region and periodically released into the drift space by a pulse applied to one of the grids. The drift space transit time is measured for various values of electric field E. The ions are detected when they penetrate the Frisch grid and induce a charge on the collector plate [2,5]. Figure 2 is a typical normal ion signal.

3. Negative Ions in He II

3.1. Normal Ion

The structure of the normal negative ion is quite well established [11]. It can be thought of as an electron entrapped within a spherical bubble, its radius at SVP being about 1.5nm. The propensity of the normal ion to nucleate, and attach to, vortex rings prevents measurements of its properties at velocities close to V_L below pressures of $\sim 10^6 \, \mathrm{Nm^{-2}}$; higher pressures suppress vortex nucleation.

3.2 Exotic Ions

If a vapour discharge is used as an ion source then under some conditions [6] a large number of so-called exotic ions are produced. The structure of these is presently the subject of speculation. Exotic ions have rather higher (low field) mobilities than, but otherwise seem to behave in essentially the same way as, normal ions - that is they nucleate vortices at velocities less than V_L [10].

3.3. Fast Ion

Although it is produced under the same conditions as the exotic ions the fast ion seems to behave in a qualitatively different manner: it does not nucleate vortex rings, even at SVP when moving at velocities greater than V_L. It seems rather likely that the fast ion is simply another member of the family of exotic ions. However, it is unique in having an insignificant vortex nucleation rate below V_L, and thus can be used to measure V_L at SVP. Figure 3 shows some typical ion signals obtained with a vapour discharge ion source.

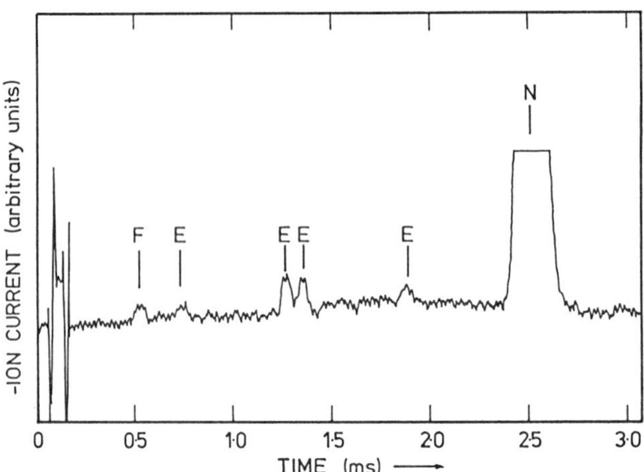

Figure 3. A typical ion current vs. time plot obtained at T=1.03K, SVP using a vapour discharge source. The fast (F), some exotic (E) and normal (N) ion peaks have been indicated. The normal ion signal has been truncated

4. Roton Emission

The influence of roton emission on the field dependence of the (average) velocity V of an ion has been studied from a theoretical standpoint by two groups. Iordanskii's calculation [12] using a strong coupling model assuming single roton emission predicts

$$V = V_L + AE^{\frac{1}{3}} \tag{1}$$

although, as discussed in [5], the model is probably not applicable to the usual experimental circumstances. In a rather simpler calculation Bowley and Sheard [13,14] argued that if an ion accelerates to velocities above V_L the number of states into which a roton can be emitted grows rapidly and can be estimated from Fermi's Golden Rule. If rotons are emitted singly then they predicted

$$V = V_L + BE^{\frac{2}{3}} \tag{2}$$

or, if rotons are emitted in pairs then

$$V = V_L + CE^{\frac{1}{3}}. \tag{3}$$

As far as the normal ion is concerned the experimental results [5] are in unequivocal agreement with (3). It would seem that single roton emission is almost completely suppressed, the reason for this rather surprising fact being unknown [14]. It is, of course, possible that a different model for single roton emission could have a result of the same form as (3). Recent fast ion results (fig. 4) are consistent with (3) but it will be necessary for further experiments at lower temperatures to be completed before we can be certain of satisfactory agreement.

It is interesting to compare the behaviour illustrated for the fast ion with that of the normal ion. Particularly striking is the small gradient of the velocity vs. $E^{\frac{1}{3}}$ graph for the fast ion. This suggests that its matrix element for roton emission must be considerably higher than is the case for the normal ion. Some care must be taken before coming to any firm conclusion for two reasons. Firstly, data for the normal ion relates to P=2.5x10^{6}Nm^{-2}, not to $P{\sim}0$ which is the case for the fast ion. Secondly,

194

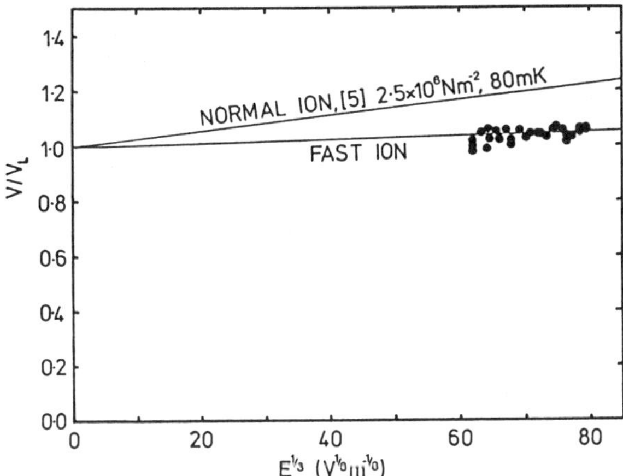

Figure 4. Ion velocity as a function of field E for the fast ion at T=1.03K, SVP compared with the normal ion data of Ellis and McClintock [5]

the effective masses of the two types of ion are not necessarily the same. However, the normal ion roton emission matrix element decreases with decreasing pressure [5] and it seems implausible that the effective masses of the fast and normal ions differ enough to invalidate the comparison.

5. Summary

There are still several questions that need to be answered before we can claim to understand properly the high velocity behaviour of ions in He II. For example, why is single roton emission apparently suppressed? What are the fast and exotic ions, and why does the fast ion not nucleate vortices?

This work was supported by the Science and Engineering Research Council of the United Kingdom.

References

1. G. W. Rayfield and F. Reif: Phys. Rev. 136A, 1194 (1964)

2. R. M. Bowley, P. V. E. McClintock, F. E. Moss, G. G. Nancolas, and P. C. E. Stamp: Phil. Trans. Roy. Soc. Lond. A307, 201 (1982)

3. J. T. Tough: In Progress in Low Temperature Physics, ed. D. F. Brewer, Vol.8, ch.3 (North-Holland, Amsterdam, 1982)

4. L. Meyer and F. Reif: Phys. Rev. 123, 727 (1961)

5. T. Ellis and P.V.E. McClintock: Phil. Trans. Roy. Soc. Lond. A315, 259 (1985)

6. C. D. H. Williams, P. C. Hendry and P. V. E. McClintock: Jap. J. Appl. Phys. Suppl. 26 − 3, 105 (1987)

7. C. S. M. Doake and P. W. Gribbon: Phys. Lett. 30A, 251 (1969)

8. G. G. Ihas and T. M. Sanders Jr. : Phys. Rev. Lett. $\underline{27}$, 383 (1971)

9. G. G. Ihas and T. M. Sanders Jr. : In Low Temperature Physics : LT13, eds. K. D. Timmerhaus, W. J. O'Sullivan and E. F. Hammel, Vol.1, p.477 (Plenum, New York, 1972)

10. V. L. Eden and P. V. E. McClintock: Phys. Lett. $\underline{102A}$, 197, (1984)

11. A. L. Fetter: In The Physics of Liquid and Solid Helium, eds. K. H. Benneman and J. B. Ketterson, Vol.1, ch.3. (Wiley, New York, 1976)

12. S. V. Iordanskii: Zh. Eksp. Teor. Fiz. $\underline{54}$, 1479 (1968)

13. R. M. Bowley and F. W. Sheard: Proc. 14th Int. Conf. on Low Temp. Phys. Vol.1, p.165 (North-Holland, Amsterdam, 1975)

14. F. W. Sheard and R. M. Bowley: Phys. Rev. $\underline{B17}$, 201 (1978)

Index of Contributors